What About China?

Answers to this and other awkward questions about climate change

Editor: Katherine Pate
Illustrator: Felix Bennett
Project Manager: Lyn Hemming
Production: Anny Mortada

First edition
Copyright © July 2008 Alastair Sawday Publishing Co. Ltd
Fragile Earth – imprint of Alastair Sawday Publishing Co. Ltd

Published in 2008
Alastair Sawday Publishing Co. Ltd
The Old Farmyard, Yanley Lane,
Long Ashton, Bristol BS41 9LR
Tel: +44 (0)1275 395430
Fax: +44 (0)1275 393388
Email: info@sawdays.co.uk or info@fragile-earth.com
Web: www.sawdays.co.uk or www.fragile-earth.com

The publishers have made every effort to ensure the accuracy of the
information in this book at the time of going to press. However, they
cannot accept any responsibility for any loss, injury or inconvenience
resulting from the use of information contained therein.

ISBN-13: 978-1-906136-11-6

Printed in Cornwall by TJ International Ltd on FSC-certified paper using
soya-based inks.

Contents

Foreword

Pat Thomas
Editor, The Ecologist

We all now know that we have to do something about climate change. Are you willing to learn a little more? You may already have a personal take on it from observing the warmer winters, the daffodils out early, the late autumns, changes in bird-life. But before we begin, let's look at the facts about climate change and global warming. You may know much of this, but stay with us – this is important.

Average global temperature has risen by 0.8°C since the start of the industrial revolution. It may not sound like much, but the consequences are enormous: the polar ice-caps are shrinking year by year, extreme weather and cyclones are on the increase, climate change is already damaging ecosystems and endangering the livelihood of millions of people. This is just the beginning.

This problem is unlike anything seen in the past. It affects the whole planet and threatens every person in every country, on every continent; a bigger threat to human life than any terrorist or world war. However, we can prevent disaster. This is not a threat coming from outer space; it is we, the people, causing climate change by – among other things – dumping too much carbon dioxide (CO_2) and other greenhouse gases into our atmosphere. This is where we have a chance to redeem ourselves! Since we started this problem ourselves perhaps we can put an end to it as well. The necessary technologies already exist.

We can – we must – cut global carbon dioxide emissions by at least 50 per cent by the year 2050. Developed, industrial countries will have to cut their emissions by as much as 80 per cent. If we do this, we can keep the rise of global temperature below 2°C; most scientists agree that this is essential if we are to prevent the climate from running completely out of control.

What is the greenhouse effect?

We need some greenhouse gases in the atmosphere to stop us from freezing. Greenhouse gases – especially carbon dioxide – trap the heat of the sun and keep the planet warm. The problem is a result of burning coal, oil and gas, leading to high concentrations of these gases in our atmosphere. As the natural balance is upset, the Earth is overheating.

We are also decreasing the natural 'sinks' (storage for gases like carbon dioxide and methane) on the surface of the planet. We seem hell bent on chopping down rainforests, removing one of the Earth's most important natural storage systems. (Deforestation contributes more than 50 per cent of the planet's carbon dioxide emissions. Trees and undergrowth are burnt on a vast scale to clear land for 'cash crops' like soya and palm oil.)

We can't carry on. Sticking to a plan of 'business as usual' will mean:
- The average global temperature will rise by as much as 5°C during this century alone.
- Flooding will increase as violent storms and heavy rainfall become more frequent.
- The world's glaciers, already melting at a frightening rate, will do so with increasing speed.
- Rivers will dry up in many parts of the world, endangering fresh water supplies.
- Sea levels could rise by several metres, threatening major cities like London, Shanghai and Tokyo – not just island states and low-lying countries like Bangladesh.
- Droughts are already becoming more frequent, for example in Africa, Asia and the Mediterranean region. Millions of people are threatened by famine, especially in the poorer countries, and this is likely to get much worse. Wealthy countries like Australia will also pay a high price for their irresponsible climate policy. Indeed, they are doing so already with severe drought and widespread destruction by fire.
- Extinctions of species will speed up, as animals, plants

and ecosystems find themselves unable to adapt to climatic changes. This danger is particularly serious for fauna and flora in coral reefs, forest, savannah, the polar regions and mountain ranges. Scientist fear that one-third of all existing species could die out by 2050.

Scientists agree that we have the technology to make industry compatible with the need to stabilise the climate. If we act now it will only cost the equivalent of one per cent of the world's total economic output, whereas doing nothing will be 20 times more expensive. Climate researchers also warn that time is running out, and that there is an urgent need to change thinking in politics and industry – and in public awareness of the problem. How we respond in the next ten years is crucial.

What now?
This book aims to dispel the myths that we all use to justify the way we live and to avoid change. It confronts the issues and ultimately the disasters that are staring us in the face.

The first step for all of us is to become better informed. This book gives a snapshot of the crises of climate change and global warming, and provides the foundation for action. Start with yourself: calculate your own carbon footprint (websites are listed in the resources), identify the quick and easy ways to cut back your own carbon consumption. Then tackle the bigger, more ingrained habits and set a longer-term goal. Finally, spread the word: tell your friends and family what you are doing and encourage them to do the same. Take it further by lobbying, campaigning and protesting, and by targeting the worst offenders in industry and business. Then go even further: join a political party or arrange a demonstration to get the message to the people at the top.

Lastly, consider this: we in the rich countries have caused most of the problems but it is the poor who will suffer the consequences first and most. Because of this, taking action on climate change is not simply a practical issue, it is a moral one.

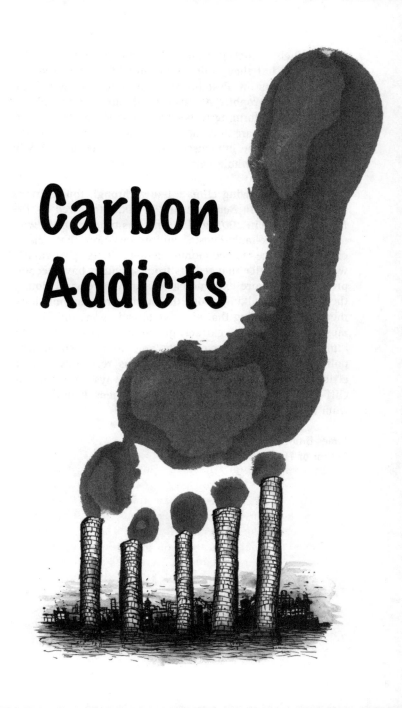

Carbon
Addicts

Carbon fuels – coal, gas and oil – underpin all aspects of modern life but they emit carbon dioxide when used and these emissions are destabilising the climate in ways that could be catastrophic. All aspects of our current culture, from industrial farming to poorly insulated houses and our means of travel, are making matters worse. But kicking the addiction could be immensely stimulating because we will have to invent a whole new way of living.

I have been following climate issues through four editions of *The Little Earth Book,* starting in 2000, followed by *The Big Earth Book* in 2007, and during this time I worked with the Irish think-tank Feasta, an organisation that relates climate science to economic issues. It is particularly worrying to notice that reports from scientists have become progressively more extreme each year. Whereas previously they were just talking about the need to reduce emissions, they are now saying that we have already gone too far and must find ways to extract greenhouse gases from the atmosphere. I have never been able to understand why politicians ignore science and drive us recklessly – with economic growth, roads and airport runways – towards the cliff-edge. Change will not come from them. It must come from us.

James Bruges
Author of *The Big Earth Book*

Q What is the point of doing anything, when China opens a new power station every week?

A "The defining challenge of our age", is how Ban Ki-moon, the United Nations General Secretary, describes climate change. It will affect all our lives, whether we take an interest or not. The biggest need is for society's climate of opinion to change.

China is making a huge effort to raise the living standards of its people. With limited oil reserves, it is turning to coal for its energy. Clean coal technologies, where the carbon is sealed underground, are expensive, but China says it will pursue this option if wealthy western nations take the lead. So far none has done so. This attitude shows the importance of leading by example: China won't do it unless our governments do it, and our governments won't do it because "it will make our industry uncompetitive". We, the electorate, must show by example that we consider the fight against global warming to be more important than commerce. Each of us is at the beginning of a chain that could influence first our own reluctant governments and then global agreements.

Three-quarters of global warming is due to the release of carbon dioxide (CO_2) when fossil fuels – coal, gas and oil – are burned. On average each person in the world is responsible for 4.6 tonnes a year. In Britain each person is responsible for 12 tonnes. A Chinese is below average at 4.2 tonnes (actually less, because many of the goods they make are exported so emissions should be counted as the responsibility of the country of destination) and an Indian is well below average at only 1.4 tonnes. An American is responsible for a whopping 20.2 tonnes (26 tonnes if you take into account the goods made abroad and imported). It would be reasonable for China to claim that its emissions per person should be allowed to rise in order to lift its struggling population out of poverty – particularly since the west has benefited historically from huge emissions over many years and is responsible for 80 per cent of the increased carbon dioxide in the atmosphere.

A change in attitude is just beginning. In the USA, due to pressure from the public, over 30 states and 600 cities have adopted policies aimed at cutting carbon emissions. We are all in this together. China is taking global warming seriously. It is phasing out incandescent light bulbs, it is building the world's most carbon-neutral city with more to follow, it has banned plastic bags in major cities, it is putting immense research into photovoltaic (PV) cells (see page 48) and other renewable technologies that convert the Sun's energy into electricity, and it is turning out thousands of graduates with expertise in these fields. Its State Council is struggling to restrain provinces and municipalities from pursuing development regardless of the effects. China's efforts to combat global warming put western governments to shame. C S Kiang, who advises the Chinese government says, "Humanity made a mistake 200 years ago and now east and west does not matter – everyone is involved. China's problems are the problems of the world. If we do not solve them together the world is going to be in a bad shape."

Human society, with its politics, world conferences, competition, economic imperatives and broken promises, could be seen as a super-tanker speeding towards the rocks and unable to stop.

But then think of those flocks of starlings you see in the evening sky: suddenly, without warning, they change direction. Our own society may suddenly change direction when strange events, or even the media, move us to a tipping point where we become alarmed that we are at the mercy of the most finely balanced and infinitely fragile of all components of this planet – the atmosphere.

Q Global warming is a natural phenomenon – look at the ice ages. Where is the evidence that human impact plays a part?

A Global temperatures have varied in step with carbon concentrations in the atmosphere, and these concentrations have suddenly shot up way above anything experienced over the past half million years. Human activity is the only explanation for this sudden surge.

Climate scientists talk about temperature, greenhouse gas concentrations and emissions. To understand what they are saying you must be prepared to study figures, think about probability and allow for scientific complexity being reduced to media simplicities.

First: temperature. There is broad consensus that global temperatures should not be allowed to rise more than 2°C above pre-industrial levels, although Professor Rajendra Pauchauri, head of the International Panel on Climate Change (IPCC) with its 2500 climate scientists, now says that 1.5°C would be more appropriate. Land-based temperatures have already risen 0.8°C, and current levels of emissions in the atmosphere commit us to a rise of 1.3°C due to the time lag between cause and effect. The biggest danger is runaway global warming. For example, if the area of arctic ice reduces there is more dark water to absorb heat from the Sun. This, in turn, causes more ice to melt – exposing more dark water – causing more melt – more dark water – more melt. This is a chain reaction that could cause temperatures to rise without any further help from us.

Second: greenhouse gas concentrations. The Stern Review (2006) stated that greenhouse gas concentrations at the time it was written were at 430ppmCO$_2$e (parts per million of carbon dioxide equivalent). A figure of 382ppm is sometimes mentioned but this does not include methane (CH$_4$) and nitrous oxide (N$_2$O). Before the industrial revolution the figure was 280ppm. According to the Stern

Review, "stabilisation at 450ppmCO$_2$e is almost out of reach, given that we are likely to reach this level in ten years". It therefore set a target for stabilising at 550ppmCO$_2$e.

Third: emissions. Despite the Kyoto Protocol (adopted in 1997), emissions have been rising at an ever-increasing rate. In 2006 James Hansen, who heads the US NASA Goddard Institute for Space Studies, said: "A global tipping point will be reached in ten years if levels of greenhouse gases are not reduced. Global warming at this point becomes unstoppable." The prestigious Tyndall Centre for Climate Change, in a 2006 report 'Living Within a Carbon Budget', said that a 90 per cent cut in UK greenhouse gas emissions, including those from shipping and airlines, is needed by 2050, adding that emissions must reduce by "an unprecedented nine per cent a year from 2010 for up to 20 years". With this sort of reduction oil refineries would no longer be viable, so we would be moving to a carbon-free economy.

What about sea levels? Greenland is the size of France and Spain combined and mostly covered with ice two kilometres (km) deep. Melt-water is dropping down crevasses and lubricating the base so that glaciers are sliding into the sea faster than anticipated. It was predicted that Greenland would lose 80km^3 of ice in 2006. NASA's Grace satellite showed that it actually lost 287km^3 that year. If greenhouse gas emissions are not stabilised within a decade, sea levels could rise several metres before the end of the century.

Q I am prepared to take risks – whenever I board an aeroplane for example. Are risks from the climate more immediate?

A Professor Rajendra Pachauri puts it starkly: "If there is no action before 2012, that's too late. What we do in the next two or three years will determine our future. This is the defining moment."

First let's look at what might happen if temperatures exceed 2°C, the broadly accepted danger zone.

The UK Meteorological Office warns that a rise above 2°C will cause havoc, with up to two-thirds of the world affected by water scarcity, major losses in agricultural productivity (grain reserves are already at a record low) and the loss of many ecosystems. With a 3°C rise the Amazon rainforest would collapse and most coral reefs would almost certainly die, the oceans would become more acidic and less able to absorb carbon dioxide. The knock-on effects from these changes are unimaginable. And if runaway warming kicks in after a tipping point is reached we can kiss goodbye to civilisation. It's that serious.

The Stern Review includes a table (see page 154) showing the likely outcomes of different concentrations of greenhouse gases in the atmosphere, and the probabilities of different disasters. Stern said that concentrations were 430ppmCO_2e in October 2006 and rising at 2ppm each year. Even if we can reduce present concentrations to 400ppmCO_2e by extracting carbon dioxide from the atmosphere, Stern's diagram shows there is a five per cent chance that temperatures would increase by almost 3°C above pre-industrial levels. In the light of the awful effects outlined above it is appropriate to ask: would you board an aeroplane if you knew it had a five per cent chance of crashing?

At 450ppmCO_2e the Stern Review gives a 50 per cent chance that temperatures will reach 2°C above pre-industrial levels, and a five per cent chance that they will climb to 3.8°C. At 550ppmCO_2e, the target adopted by the Review, the diagram shows a 50 per cent chance that temperatures will reach 3°C and a five per cent chance that they could soar to 4.7°C above pre-industrial levels.

Sir Nicholas Stern is an environmental economist. His discipline requires him to assume that our economy is sacrosanct and to explore what it can afford to spend on environmental issues.

Ecological economists invert the question. They see the human economy as just one component within the planet's ecology and ask what limits are imposed on economic activity by the environment. They say that you cannot extrapolate from past events because natural systems frequently tip from one stable state into another. We already over-exploit the world's resources, and all our efforts should now go into repairing the damage and learning to live within the Earth's systems before it is too late.

Q It's the responsibility of politicians and world leaders, not me. What difference can I make?

A Let's imagine: a parent and child are watering their garden. Father controls the tap while daughter holds the hose. When they've finished the job he says, "That's enough now, stop all the holes in the spray head with matchsticks". "Dad", she replies, "are you mad? Turn the tap off!"

The father's stupid approach is rather like present policies where we are urged to use low-energy light bulbs and travel less while politicians and world leaders allow, even encourage, the extraction of as much fossil fuel as possible. Can't they understand that, once out of the ground, these fuels will be burned and the carbon dioxide they release will reach the atmosphere? If politicians are serious about greenhouse gas emissions they must find a way to reduce, not increase, the amount of coal, gas and oil that is extracted within, or imported into, a country. Unless they control the tap all our attempts to reduce emissions in a hundred little ways will be useless.

But how will politicians be persuaded? Unless the public puts pressure on them corporate interests will rule.

Corporate lobbyists will twist any complicated legislation to their own advantage. Politicians pretend to the electorate that they are concerned when they introduce targets, taxes and incentives, but these blunt instruments cannot guarantee that their commitments will be met.

A new wave of thinking suggests that the solution must involve individuals. The atmosphere does not belong to corporations – not even to countries or governments. We all, as individuals, have an equal right to its life-maintaining properties. This new thinking led to the suggestion that everyone should have a personal carbon allocation managed with the help of a smart-card. Points would be deducted every time you filled your car or paid your heating bill. And you could sell any surplus points so that the gas-guzzler would subsidise those with a low carbon footprint. This approach would have a huge psychological impact since it would make us all aware of our responsibility for carbon emissions. But it would be very difficult to administer even in an industrialised country.

For a policy to be adopted it must be simple, and for it to be politically sustainable it must be popular with the electorate. So the approach has been modified as Cap-and-Share – in America a very similar policy is called the Sky Trust – which would be easy to introduce. Under Cap-and-Share much of the astronomical income enjoyed at present by fuel producers would go to individuals. The Irish government may be the first to adopt it. This is how it could work for Ireland's transport sector:

- A cap, which reduces each year, sets the maximum emissions allowed from all road transport.
- All adults receive equal emission-permits that, together, add up to the amount of this cap.
- People can then sell their emission-permits through brokers to whichever fuel-importing company offers the best price.
- Importers would not be allowed to sell fuel unless they had enough permits to cover its emissions.

With less fossil fuel coming into the economy, its price would rise and push up the cost of living but, instead of the oil companies making ever-larger profits, people would receive an income from the sale of their permits to compensate for the higher prices, and local economies would flourish because of this injection of money at grass-roots level. The government would use its normal powers to ensure that essential users, like ambulances, received their necessary share, and that biofuels were not allowed to compete with food production.

The Irish application of Cap-and-Share to an individual sector of the economy demonstrates that it could run alongside the European Emissions Trading Scheme (EU-ETS) – until the ETS collapses. If Cap-and-Share were introduced on a global scale, or even within a large country like China or India that is struggling with discontent as world food prices rise, much of the money that at present pours into oil-rich countries would stay and automatically relieve their rural poverty.

What difference can you make? Urge your politicians to put a cap on the use of fossil fuels and share the right to benefits from the use of those fuels equally among all citizens. And join the transition movement where communities work together to find a post-carbon way of living.

Q Governments are encouraging us to buy green electricity and install renewable energy units. Isn't that enough?

A Every little helps, but there is an interesting story behind this one. People living in Germany have much more incentive to reduce emissions than people living in Britain.

Germany introduced the Feed-In-Tariff (FIT) for the supply of electricity. There are three parts to this system, all of which are essential to its success:

1. You have a right to feed any electricity you generate into the national grid and the power companies are required to accept it.
2. You are paid four or five times the tariff for the electricity you generate for a guaranteed period of 20 years.
3. There is no limit on the amount of electricity you can feed in.

With this system you can recoup the cost of installing PV cells or panels, or wind turbines in a short period, and subsequently receive a significant income. You are, in effect, providing part of a large dispersed power station that belongs to individuals and communities.

Power companies had enjoyed control of energy and, not surprisingly, were furious to see this taken from them and given to citizens. They challenged the policy in the Bundestag (the German parliament), they fought it in the courts and they took it to the European Court on the grounds that setting guaranteed prices infringed Europe's free-market policies. They lost each time.

Individuals, farmers and communities in Germany are now enthusiastically erecting wind turbines and covering their roofs with PV cells. The normal electricity tariff is set high enough to enable energy companies to buy electricity generated by individuals and communities at the higher rate set by the FIT. In this way the scheme is self-financing, requiring no expenditure by the government. But, even so, the price of electricity is lower in Germany than in Britain.

Germany now has 20 times more renewable energy than the UK, though it started later and has less wind. The advantages have been demonstrated; the system is spreading around the world and Germany is at the heart of the new renewable-energy industry.

And what a pension plan! Why put money into PEPs or stocks and shares, when you could invest in PV cells or wind power that have guaranteed returns? Money increased in value when it could buy as much energy as was needed, but those days are over. Money savings, as we are seeing, can collapse. In future, energy will be the scarce commodity that will always be in demand. Energy, not money, is bound to increase in value and give you a secure income.

The UK made the mistake of consulting energy experts when it introduced the bureaucratic 'renewable obligation'. Energy companies and their experts, who have until now enjoyed a monopoly of the supply of electricity, want generation centralised even though they know that a centralised system wastes three-quarters of its power through inefficient generation, transmission and end use. They make difficulties over connections. If the government requires developers to build carbon-neutral schemes they lobby for them to be allowed to purchase offsets from the supplier instead of being genuinely carbon neutral. They want huge arrays of wind turbines far out to sea, which only they can build. They know that carbon fuels have a limited life so their long-term objective is not for gas and coal-fired power stations but to perpetuate nuclear power. Uranium is a limited resource so they want the industry to be propped up until nuclear fusion (see page 52) can be introduced. With fusion, as opposed to fission, only one or two plants would be needed to generate enough electricity for the whole country: an ultra-centralised system that would provide ideal targets for terrorists and put an end to democracy.

The UK, to our shame, has been singled out by the United Nations (UN) as the European country making least progress in the introduction of renewable energy supplies.

Q OK, we're all in this together so what else can we do? Can we capture carbon dioxide that's already in the atmosphere?

A It's easy. Plants, through photosynthesis, capture carbon dioxide all the time. They release it when they decompose. Cut the plants, turn them into charcoal – sometimes called biochar – before they decompose and bury the charcoal. It's like coal mining in reverse.

In India, Ravi Kumar has developed a domestic 'pyrolysis unit' that is suitable for home use. A family collects waste plants – even grass cuttings – dries them, and puts them into the unit's circular casing. The casing is sealed so no oxygen combines with the plants' carbon to make carbon dioxide. To start with, a few sticks are burned in the central void. When the plants in the surrounding casing heat up they emit gases through small holes into this void. The gases ignite and continue to burn for over an hour, long enough to cook a meal. Finally the charred plants are raked out.

But that's only a part of the story; the best part is to follow. Charcoal in the ground attracts microbes and nutrients and the soil's fertility can increase spectacularly. And, what's more, the soil's subsequent ability to capture and retain carbon dioxide also increases.

There is one small problem. To start with the charcoal attracts nutrients from the surrounding soil, so the soil's fertility may drop for a time. The solution is to prime it before distribution, and this can involve the sewerage system. Indian eco-san toilets separate urine for immediate use and guide faeces into two chambers that can be emptied in turn after six months. Both urine and matured faeces, together with farm manure and compost, can be added to a charcoal pit before being spread onto the land. This system captures carbon, increases soil fertility, then captures more carbon and revolutionises the sewerage system. A win-win-win-win solution. If it were widely used, India would need no chemical fertilisers.

Forget about important people meeting at important conferences, emitting hot air from their mouths and greenhouse gases from their flights. Forget about mega-scale technofixes. Think small-scale multiplied by a billion ordinary people throughout the world. If rural India can do it, we could also have pyrolysis units for our back gardens or allotments using waste plant material and wood from skips. Perhaps we could convert our garages into workshops and build the units ourselves. This is a technology in its infancy so there is plenty of scope for experiment, community co-operation and business start-ups. A lot of research is now in progress to find what plants make the best charcoal for soil fertility and which kinds of soil benefit most.

Big farms will need the more sophisticated units that are being developed by western industries. But there is a danger. These companies want the sequestration of carbon dioxide to be awarded carbon credits in order to finance their development. The incentive to sequester carbon dioxide would then be through globalised neoliberal trading, unrelated to farming. There are bound to be unexpected consequences. Small-scale farmers could not possibly enter into carbon credit contracts, they would not be able to measure results, and monitoring would require an army of – possibly corrupt – bureaucrats. Agri-businesses would be the only organisations able to benefit. As the value of carbon credits rose, land would be used for the most efficient sequestration-crops in competition with food-crops, and small-scale farmers would be out-competed and join the disastrous migration to city slums.

Charcoal-enriched farming is not new. Large areas of 'terra preta' – rich dark fertile soil – have been found in the Amazon rainforest where surrounding areas have hardly any topsoil. Pre-Columbian pottery shards indicate that a civilisation once thrived on this technique. Fungal activity has perpetuated the soil's fertility ever since.

The importance of sequestration using charcoal cannot be overstated. Excess greenhouse gas concentrations in the atmosphere have created a global emergency. We must drastically reduce emissions but, in addition, the Charcoal Revolution may be the best way to extract carbon dioxide and get us back to stable pre-industrial concentrations. To achieve this, the sequestration of carbon dioxide using charcoal could become standard farming practice throughout the world.

Q How will the economy cope if we prevent our industries from using coal, gas and oil?

A The country or company that persists with yesterday's technologies has no hope. A changing climate requires fundamental change to our way of living and to the way we do business. Change is a challenge and an opportunity.

Experts have always told us that nothing can be changed: trade rules that destroy the livelihoods of billions around the world; the money system that creates such obscene inequality; the obsession with commercial competitiveness; retail practices that make millions for the supermarkets and drive those that actually grow our food into penury; gender-bending chemicals used in plastics, manufacturing and food production; the dictatorial power of corporations; global political power in the hands of countries that invade others and prosper from the arms trade. These things are all driving climate change. The dinosaurs that run our countries must wake up or be replaced – change or die. Change is an imperative, therefore change is possible. We can now create a new society.

But even if we lower our sights to the way business managers think, new opportunities are on offer. Scribes weren't trained once printing was invented; yew trees weren't planted once gunpowder took over from longbows.

The crude fossil fuel technologies – heat, beat and treat – served the industrial revolution and still survive as the backbone of today's economy. But they have no future. We use vast amounts of fossil fuels to achieve high temperatures and pressures for making new forms of ceramics, but we can't better an oyster for sheer strength or for effective underwater glue. A spider's web is stronger, size for size, than any steel we can make. Agricultural science, for 50 years, has been based on using natural gas and oil for fertilisers, pesticides, water pumps and tractors, and it is only recently that scientists have turned their attention to the way nature does it; they call it biomimicry and biologists are becoming team leaders in many businesses.

A country that insists on maintaining yesterday's technologies for its commercial viability will decline. They should only be used to make the transition to carbon-free technologies. When Denmark pioneered the new industry of wind turbines it cornered half the world's new installation and is now receiving huge orders from China. Germany's enthusiasm for renewable energy has created 250,000 jobs in the last six years. The UK's reluctance has generated a paltry 25,000 jobs.

Half of the energy we use is devoted to producing economic growth. Since endless growth is not possible on a finite planet, this is a gigantic waste of energy. Other aspects of our economy are equally nonsensical. For example, in 2007 the UK imported 14,000 tonnes of chocolate-covered waffles and exported 15,000 tonnes of the same. In the same year the UK exported 20 tonnes of mineral water to Australia; Australia exported 21 tonnes of mineral water to the UK. Both the imports and exports added to the country's gross domestic product (GDP) by which its health is gauged. Could anything be more ludicrous?

Global free trade, the mantra that drives western politicians (see pages 96 and 98), means that resources, goods and food can be imported from whichever country is prepared to sell them cheapest. This destroys investment in industry and agriculture in wealthier countries and makes the global economy dependent on fuels that have little future. Rising countries like China and India take advantage of this strange mantra to secure resources for themselves that will in due course be unavailable to others.

Our mad policies – trying to compete in the global economy with yesterday's technology – will produce a long and deep recession that may give us time to reconsider our approach. Our economy will be healthier when we adopt sane trading policies and invest in tomorrow's carbon-free – not yesterday's fossilised – technologies.

Q **Don't we need growth to maintain our jobs and way of life? Even if this harms the planet, a no-growth economy is just not possible.**

A Sadly this is true – but only because of the money system we use.

Did you know that the British government only makes five per cent of the nation's money? It has allowed the rest to be created by private businesses for profit. The more money they can create, the more profit they can extract. This is why US banks encouraged people to borrow more than the value of the houses they were buying – known as the 'sub-prime' mortgages – in the hope that their houses would increase in value, a disgraceful practice purely to promote economic growth. House values did not increase. People lost their houses, their savings and, because this caused a credit crunch, their jobs.

Ninety-five per cent of all money is made not by government but by banks and building societies putting you and me into debt so that we can pay them interest. When you walk into a bank and take out a loan the manager doesn't check that he or she has the money in the bank's vault, they simply tap figures into a computer and, abracadabra, the money suddenly exists.

The money they create is virtual money. Why? Try this trick on your bank manager. Ask for a loan to start a business, offering your house as security. She will tap into a computer and the money is yours to use. Then, before you leave, say sorry, you don't want the money after all. She will say that's no problem, tap into her computer and the money vanishes. The money existed, was 'in circulation' for ten minutes, then abruptly ceased to exist. Only the money the government creates is real and a bank only has to have a tiny proportion of the money it lends as real money. That's why the UK bank Northern Rock was unable to pay its investors real money when they demanded it *en masse*. Most of the money it was storing and lending as mortgages was unreal – virtual, fictitious – money.

On a mortgage of £100,000 you may eventually pay an additional £100,000 in interest to support the lifestyle of the bank's directors. The economy needs to grow if interest is to be paid on top of the original loan. But if loans were to attract no interest these private institutions would not bother to create any money to lend. There would be no money in the economy – no economy. So a stable economy, one without growth, is not possible with our present money system. The alternative to a growth economy is not a stable economy but a collapsed economy.

A stable economy would only be possible if private institutions – banks and building societies – were forbidden to create money, just as you and I would end up in jail if we were to counterfeit bank notes. An issuing authority should create all the money that is needed for exchange, without charging interest. The issuing authority should be independent of government so that the creation of excess money cannot be used to boost spending before an election, but its members should be elected. We would then pay banks a fee for any services they render, just as we pay accountants or engineers a fee.

The owner of a skyscraper does not like to be told that the building's foundation is faulty. Similarly, politicians don't like to be told that the foundation of the economy – the money system – requires unsustainable growth and is therefore faulty. They claim to decouple growth from the use of resources, but this only means that we provide services – that require no resources – and import our manufactured goods instead of making them ourselves. Europe's trade deficit with China in 2007, for example, was £90 billion, so all these goods made in China should count against Europe's account for resource-use.

Environmentalists say that we should aim at growth in the quality of life; pointing out that while the economy has grown massively, the quality of community and social life has been deteriorating in recent years. They are right, but there is little money in this to buck the imperative of the money system. Bankers have huge influence over government policy so they will sit tight on the amazing privilege granted them – to create the nation's money and extract interest – until their fantastical house of cards collapses around them.

Western economies will continue to destroy the planet's support systems as long as their economies are based on growth.

Q Isn't global warming caused by Sun activity? Summers were much hotter when I was a kid.

A The delicate balance of the Sun's intensity is critical to maintaining life on Earth. Sunspot activity causes some warming from time to time and cosmic rays affect the clouds. Scientists have factored these into their computer models for global warming.

The Earth's distance from the Sun varies on a 100,000-year cycle and kickstarts interglacial periods where increased carbon dioxide warms the Earth. We are currently moving further from the Sun, so we are due for another cooling, but we have put so much carbon dioxide in the atmosphere that this cooling may not happen.

The Little Ice Age – on and off between the 14th and 19th centuries, with minimum temperatures in the 16th century when the Thames sometimes froze over – was confined to the northern hemisphere and may have been caused by solar activity or dust from volcanoes.

These variations in global temperature are hardly surprising when you consider that the atmosphere is so thin – it has been likened to the outer skin of an onion – and is the only thing that protects us from the frozen darkness of space. The most amazing statistic is that the Sun has increased in intensity by about 30 per cent since life began on Earth, but the temperature close to the ground has remained fairly stable due to the interaction of plants with the atmosphere. Human activity now uses a large proportion of the biological activity on the planet, so it is not surprising that the waste exhausts from our frenetic activity are having an effect on the fragile skin of the atmosphere.

Exhausts may not only come from us. In Siberia you can scrape away 30 centimetres of snow, drive a pipe into the permafrost and light a match. A jet of flame will rush upwards. This is from methane, a greenhouse gas 20 times as potent as carbon dioxide, waiting to be released if temperatures rise sufficiently to cause the permafrost to melt. It is a time bomb waiting to explode.

Will natural systems be able to correct the changes that we are making to the atmosphere? The answer depends on whether we carry on with business as usual. If we do there is no chance that natural corrections will be adequate to keep the climate tolerable for our species. But if carbon emissions are dramatically reduced within ten years and if we are able to extract some of the atmospheric carbon dioxide, natural systems may be able to repair the damage caused by the greenhouse gases we have already released.

Q Why bother? Erupting volcanoes emit far more gas into the atmosphere than we do.

A In 1883 the volcano Krakatoa in Indonesia, exploded with about 13,000 times the energy of the Hiroshima atomic bomb. The blast was heard 3500km away in Australia. Ash spewed into the sky to an altitude of 80km, circling the globe and causing chaotic weather patterns and spectacular sunsets for many years. Cirrus clouds received an injection of sulphuric acid and were spread by high-level winds. The sulphur made these clouds more reflective, so the Sun's rays were partially reflected back into space. The Earth cooled 1.2°C and only returned to normal after five years, when the sulphur had fallen to earth as acid rain.

The Krakatoa gases disturbed the climate but only lasted in the atmosphere for a few years. Carbon dioxide lasts a lot longer – several hundred years – before it is absorbed back into the earth. Another difference is that the Krakatoa gases reflected sunlight, causing cooling, but carbon dioxide traps solar heat, leading to warming.

The climate naturally tends towards a stable state; when disturbed it gradually returns to stability. It's like a ball being kicked about in sand dunes usually rolling back into the hollow where you stand (as with the Krakatoa eruption). But the ball might go over a ridge and roll into a different hollow. The last glacial period was one of these hollows, and it excluded humans from much of the planet. We don't know for certain how far we have kicked the climate. It could be balancing on a ridge where another jolt will roll it into a different hollow. If this hollow allows runaway global warming we will be in deep trouble.

Volcanoes also expel carbon dioxide. The initial injection of carbon into the atmosphere of the young Earth came from volcanic activity, and helped to produce a warm and stable climate conducive to life. But volcanic activity now releases less than one per cent of the amount released by human activities.

Rising sea levels caused by climate change may result in more volcanic activity. Seismic faults tend to run close to coasts – such as those of California, Indonesia and New Zealand – and the increased weight of water on one side of a fault may precipitate earthquakes or eruptions.

Volcanoes in Siberia emitted carbon dioxide and caused some warming 350 million years ago. This melted permafrost, which in turn released methane to result in runaway global warming. Ocean currents stalled so that deep water was deprived of oxygen and produced clouds of hydrogen sulphide that led to mass extinctions on land. 90 per cent of marine life and 70 per cent of terrestrial life were wiped out. The planet took ten million years to recover.

The two previous mass extinctions were also caused by volcanic activity that resulted in runaway global warming. Only the Cretaceous extinction, that did for the dinosaurs 64 million years ago, was caused by the impact of a meteorite.

It is essential that we stop business-as-usual – the commercial imperative beloved of our governments with their massive bureaucratic bodies, short legs that trample the weak, long necks that allow them to keep a check on us all, and small brain boxes that can handle only the lowest common denominator of thought – kicking the climate over a ridge into runaway global warming.

Q Why worry? Technology will sort it out, won't it?

A Greenhouse gases from burning fossil fuels move slowly upwards into the sky, where they stay for a time until they come slowly down to be absorbed by plants and oceans. Emissions must be reduced and carbon must be captured in order to reduce the volume of greenhouse gases in the atmosphere.

Emissions will be reduced if we develop a low-carbon culture. This links the use of technology (say, low-energy LED light bulbs) with lifestyle changes (say, travel by train rather than air). The other chapters in this book show ways that each of us can live and use technology appropriately to reduce our carbon footprint.

Governments must also be more sensible. For example, three-quarters of the electricity generated in the UK is wasted through inefficient generation, transmission and end use, as a result of the government policy of leaving decisions to the market. Big remote power stations – with cooling towers that symbolise the waste – were once the cheapest way. For 60 years Sweden has been much more sensible with its policy for combined heat and power. This requires energy companies to build small power stations within towns so that the heat resulting from electricity generation can be used to heat houses and offices.

Technology has increased the efficiency with which energy is used, particularly in wealthy countries, but we need to beware of Jevons' paradox: greater efficiency has always resulted in an increase, not a decrease, in the use of a resource.

Beware of big-scale new technologies – they have had a fearful history. The industrial revolution transformed the world, but it failed to prevent the unrelenting tragedy of poverty in the developing world and it introduced global warming, the greatest threat humanity has ever faced.

We were told that nuclear power would provide carbon-free electricity too cheap to meter, but it led to 45 years of Cold War and is again a major cause of conflict between nations. There is today the threat of terrorist attack on a nuclear power station that could result in unprecedented devastation; and also the significant costs involved with building nuclear power stations, the problem of the disposal of hazardous waste products that result from the process, and the cost of any reprocessing. It is another non-renewable technology because uranium mines have passed their peak and when new reactors come on-stream demand for uranium will exceed supply. The price, as with oil, will soar and conflict will erupt to secure the uranium available. The industry now promotes nuclear fusion (see page 52), as opposed to fission. But, if all goes well – and given the history of nuclear, this is a big 'if' – the first electricity from a fusion plant will not reach the grid for 50 years. Sceptics see the pilot plant in France, ITER, as an exciting job-creation project for plasma-physicists.

Plankton in the oceans absorb some carbon dioxide. Most of it finds its way back into the air but some is absorbed by water and some sinks to the ocean floor where a small proportion is trapped. Some companies suggest fertilising the surface water to grow more plankton, using either iron filings or urea, or by drawing nutrients up from the sea-bed. Set against this is research showing that, with increased carbon dioxide from the air, the oceans are becoming more acidic and krill (shrimp-like creatures at the base of the food chain) may become unable to grow protective shells. By adding further carbon dioxide to the oceans the whole marine ecology could collapse.

There are many exciting ideas for introducing large-scale technology that will allow us to carry on polluting, but most of them require fundamental changes to the atmosphere or the oceans – elements on which all life depends – so the unexpected consequences could be terminal.

The only safe technologies are those we can control on land and those that help return the oceans and the atmosphere to their pre-industrial state.

Q Bring it on! Global warming means that we'll get great summers, why should we change our lifestyles?

A That's a nice idea but, unfortunately, not realistic even if you are thinking in the short term. Our interference with the climate will have all sorts of unforeseen consequences and it would be a miracle if they all turn out to be pleasant. We hoped for a long hot summer in 2007 but hardly had a summer at all. Greece, Turkey and California, on the other hand, had heatwaves and catastrophic fires.

In 2007 the unpredictable weather caused an unprecedented reduction in harvests of most staple food crops and world grain reserves are at a record low. The UN reports that, due to climate change, 34 countries are already facing long and sustained food shortages. In a misguided attempt to replace fossil fuels, crops are being diverted to feed cars rather than people; and the cars of the rich are able to pay much more than destitute slum-dwellers.

Our western lifestyle is wonderful for those who can afford it, pay interest on a mortgage and fly far away from their neighbours whenever they have a few days' leisure. It has been built up with total dependence on the fuels that cause global warming. But is it so wonderful? Even those who enjoy the benefits are subject to long hours of stressful work and anxiety about the future. Others – most of the world's population – are excluded from this consumerist nirvana.

Coal started the industrial revolution, and oil provided the most easily used form of energy that has ever existed. We will never again have such a bonanza. When the use of fossil fuel is drastically curtailed, as it will have to be to prevent climate chaos, our western lifestyle will have to change dramatically.

We live in exciting times because we now need to invent a whole new low-energy economy and lifestyle. We can either regard this as a disaster or treat it as a life-enhancing challenge. We will need to build stronger communities, we will need to develop a wholly new way to source food, and prospects will open up for our initiative and imagination to meet the challenge.

Q Should we go back to coal? I have heard that they can now capture the carbon and bury it.

A Countries are returning to coal in a big way because the production of oil is not keeping up with demand and its price is escalating. China, India and the US are short of oil but have large reserves of coal.

This is slightly odd, since it indicates that humanity has a remarkable ability not to learn from experience. Coal, like oil, is a non-renewable resource so will inevitably, at some stage, have the same problems as oil depletion. Some analysts believe that the peak in coal extraction is not far away. Coal is the dirtiest of the fossil fuels and emits more carbon dioxide per kilowatt–hour (kWh) of electricity generated than oil, and twice as much as natural gas. The return to coal will lead to catastrophe if unchecked.

There is a lot of talk about 'clean coal' or CCS (carbon capture and storage), where the coal's carbon emissions are trapped and pumped into a deep geological seam. As yet, this is an untested technology. The process requires additional coal to be mined, not all emissions are captured and there are emissions due to mining and transport, so it will only be about 75 per cent efficient. Compared with other fuels, CCS coal would release half the emissions of local natural gas, and maybe the same emissions as liquefied natural gas. So clean coal should really be called 'carbon-light coal'.

Carbon-light coal fails dismally when compared with wind power. Its carbon intensity is 320 grams (g) per kWh whereas that of wind is 5g per kWh. Also its energy return on investment is only 2.5 times, compared with wind's 34 times.

The real drawback is that CCS technology will not be widely available for another 20 years at least and during this time a lot of carbon-intensive energy will be used to install the technology. Remember, scientists say emissions must drop significantly within ten years to avoid climate catastrophe. CCS, therefore, is counterproductive in fighting the present climate crisis.

CCS is being used as a public relations exercise for allowing coal-fired power stations to go ahead during the next 12 years. The UK government says that new plants will be 'CCS-ready', which means that CCS can be added at some time in the future – if it ever becomes viable. There is clearly a view among politicians that 'grown-ups' must face up to tough, big, dirty and expensive solutions. Clean, efficient, long-term and renewable solutions that are immediately available are beneath their dignity – for children to dream about. These naive children, however, want a future.

Q **Isn't climate change nature's way of keeping the population under control? The planet isn't big enough to sustain all the people on it.**

A Most species, if given excessively favourable conditions, multiply to such an extent that they exhaust the resources on which they depend and suddenly collapse. Think of locusts. It is through evolution that the balance of nature has been achieved and maintained. The human population has been growing exponentially but global warming, among other things, may bring a correction.

Pathogens like malaria are moving with the changing climate. Bluetongue disease has been spreading northwards since 1998. It reached a cow in England in September 2007, has since spread to sheep and no vaccine has yet been developed for the northern serotype. This is just one example of the way global warming might affect our source of food, though luckily this one doesn't harm humans.

Poor countries are suffering the effects of climate change more than rich countries – particularly floods and droughts. The proportion of poor people in the world is rising, due to their higher fertility rate. In 1950, globally, there were two poor for every rich person; today there are about four. If the trend continues, in 2025 there will be about six poor for every rich person. These trends point to a future of massive economic migration. Walls are going up. Israel has built a wall; the US is building a fence to keep out Mexicans; Britain has a moat; and Europe might re-introduce a wall to its east.

Our species likes to overcome problems by applying intelligence, but it often upsets the balance of nature in the process. Antibiotics, for example, have had amazing benefits for human health since the 1940s but bacteria are now developing resistance to them. Bacteria are held in check by competition with other bacteria, so continued use of an antibiotic after most bacteria have been killed leaves the field open to the resistant bacteria to multiply unchecked. This is already being experienced with MRSA (methicillin-resistant *Staphylococcus aureus*) in hospitals. With farmed animals it is even worse. The intensive rearing of animals in gigantic sheds provides ideal conditions for resistant bacteria to breed, yet until recently farmers added antibiotics to the feed, to make livestock grow faster and resist diseases incurred by their horrendously unnatural conditions. Although using antibiotics as a growth promoter is now banned, antibiotics are still widely used as medication (see page 78). New strains of MRSA are resulting from the use of antibiotics in pig production and these are spreading to other animals and to humans. Antibiotic-resistant *E. coli* in chickens is spreading among intensive animal farms; this is particularly worrying because it is similar to a strain that affects 30,000 people each year in the UK (see page 77).

We have disturbed the balance of nature, and climate change may unbalance it further.

Q Floods and rising seas can be held back with defences, can't they? Anyway it's all a part of the Earth's natural cycle; the British Isles used to be joined to Europe, didn't they?

A In the last glacial period, when more of the Earth's water was locked in glaciers, sea levels were considerably lower than now and we wouldn't have needed a channel tunnel. Ice was so deep over much of the UK that its movement gouged out the valleys we see in our landscape today. Then the global climate warmed, the ice melted and sea levels rose, so water covered the land that once connected Dover to Calais.

The global climate is warming again. Glaciers slipping off Greenland could raise sea levels significantly this century.

Antarctica, 28 times the size of France, is covered by ice around 2km thick. Scientists were surprised when the Antarctic Larson B ice shelf disintegrated in 2002. It was floating on water, so its melting did not affect sea levels, but it also held back land-based glaciers which are now moving seawards faster than before.

Ten years ago it was thought that sea levels would rise less than a metre by 2100, but research is now showing that they could rise between two and four metres this century, and eventually 15 to 20 metres or even more if runaway global warming were to kick in.

Stand on the Victoria Embankment in London and ask yourself if barriers are a practical option. Think of the many coastal nuclear plants. Consider the low-lying parts of the UK that are only just above sea level, and The Netherlands, which is largely below sea level. The fate of New Orleans in 2005 was a graphic illustration of what happens when flood defences fail.

What is happening is the Earth's natural response to excessive carbon dioxide being pumped into the atmosphere.

Q Isn't carbon offsetting all I need to do to justify flying?

A The carbon offset theory is that we can fund projects in other countries to reduce carbon emissions and thereby claim that we are not harming the atmosphere through causing emissions ourselves. It swaps emissions in one place for savings in another, so has no net value in fighting climate change.

Carbon offsetting has become big business. Companies vie with each other to claim most emission-reductions for least money. Tree planting is one way, but it takes many years for a tree to absorb the carbon dioxide it is claimed to be offsetting and, by the end of its life, a tree emits as much as it has absorbed. Scientists now say that even during its early growth a tree's performance is questionable. Planted in a poor country, trees may prevent local farmers using the land for cultivation, so offsetting can be regarded as a form of colonialism – their land serves our energy needs.

In South Africa one offset company delivered thousands of low-energy light bulbs to a rural area and sold carbon credits to its customers in the UK on the grounds that their emissions would, in effect, be cancelled by reductions elsewhere. It was not around to replace breakages and the villagers could not afford replacements. As it turned out, the local electricity company had been planning to distribute and maintain similar bulbs in the same area in order to avoid having to construct a new generator. The 'offsets' therefore resulted in more emissions than would otherwise have taken place.

A better approach is demonstrated by Social Change and Development (SCAD), an organisation in south India that has for years been helping villagers plant trees to retain soil and improve the micro-climate as well as for their fruit, fuel and compost. SCAD provides advice, materials and incentives for maintenance. Whether or not the trees effectively capture carbon, this would be a good recipient of conscience-money. SCAD is also working with a UK organisation, Converging World, to install wind turbines.

Instead of paying money to pretend that the emissions we make today will eventually be cancelled by reduced emissions elsewhere, we could take steps to reduce our carbon footprints. Here are some suggestions. Eat locally grown organic food (vast amounts of fossil fuel are involved in conventional farming), insulate your house, travel by train rather than plane, install PV roof panels that convert sunshine into electricity and, of course, bury charged charcoal in your vegetable patch (see page 19). However, in the present crisis we need to do more than that – we need take carbon out of the air.

Carbon offsetting is rather like medieval indulgences where, by making payments to the church, you could have your sins forgiven and buy a place in heaven.

Q Isn't carbon trading a con to enable governments and big polluters to continue with business as usual?

A How could you suggest such a thing? Since the Kyoto meeting in 1997, world negotiators have spent 12 years flying around the world and holding serious meetings with all the bureaucrats available to negotiate sophisticated mechanisms – joint implementation, clean development and the like – to reduce carbon emissions.

These trades enable low-carbon projects in poor countries to be funded by businesses in rich countries. The poor want to encourage the trade because they receive development. Companies are happy to contribute because they receive carbon credits that allow them to make more emissions than would otherwise be allowed under the Kyoto Protocol.

Does this cap-and-trade policy benefit the climate? At best – fewer emissions in one place but more in another – the trade is carbon-neutral and does nothing actually to reduce emissions unless the cap can be enforced worldwide. The low-carbon plants in poor countries may result in fewer emissions than would otherwise be the case, but the additional development this generates invariably results in more. Anyone familiar with poor countries will be aware of low-carbon projects shelved in the hope that, at some time in the future, they might attract carbon credits; these credits can therefore be responsible for delaying emission-reducing schemes while allowing increased emissions in industrialised countries. It is more realistic to see the trade in carbon credits as scripted by vested interests looking for ways to carry on with business as usual. Since these mechanisms were introduced, global emissions have been rising at an increasing rate.

Europe has introduced its own emissions trading scheme to try and contain the rise in emissions. But the EU-ETS has been a disaster. The public couldn't understand what was going on so it was left to big business to influence the negotiators. Unbelievably, emission permits were given, not even sold, to the biggest polluters! A study in 2006 by The Energy Research Centre of The Netherlands, found that power companies got a windfall profit of €1.5 billion in 2005 alone. A World Wide Fund for Nature paper calculated that German utilities will get windfall profits of between €31 and €64 billion during the life of the scheme. Even the aviation industry will receive about €3 billion as a gift. Meanwhile users of energy, including the public, have paid the cost of these permits. And EU-ETS has not been responsible for any reduction in Europe's emissions.

If emissions must be drastically reduced within a decade, as scientists state, then measures must be introduced to achieve reductions without complicating other aspects of a country's economic life. Carbon-trading schemes are complex, take years to set up, have to be carefully monitored and are often counterproductive. Their achievement has been to allow business-as-usual to continue unchecked – which was the objective of the corporate lobby that helped them into being.

Q **I will be long gone before anything bad happens, won't I?**

A Scientific journals constantly report climate events happening earlier and more dramatically than previously predicted. They also mention the possibility of tipping points, where the climate could change into a totally new state. They don't know how close we are to such tipping points.

In 2005 the UN issued four global appeals for disaster aid – a record, although only two of them were caused by climate events. In 2007, by October, the UN had issued 13 global appeals, 12 of which were climate-related and devastated the lives of maybe 100 million people in Africa and Asia. As conditions deteriorate in many poor countries, the number of migrants fleeing starvation is set to explode.

Insurance companies struggle to compete with each other by keeping prices down, but have to take a hard-headed assessment of risk. They are increasingly raising prices and excluding categories subject to climate change.

The price of oil is escalating and will rise further if effective emission curbs are put in place. This is already affecting the cost of living because almost all of our food and gadgets are dependent on the availability of cheap energy from fossil fuels.

Environmental issues are in the media every day. In France 15,000 people died during the heatwave in 2003 and now there is a daily slot in the main news broadcasts exclusively devoted to the environment. In 2007 Britain suffered from widespread floods and California suffered the worst fires on record.

We should not blame every individual weather event on climate change. There have been many unusual occurrences in the past, but it is the pattern of increase of these events that points to the effects of a warming climate. Land temperatures are still only 0.8°C above pre-industrial temperatures. There is every chance that this figure will double, with more than double the effects, and it may go significantly higher.

It has been appreciated for a long time that a changing climate would not be pleasant, and the younger you are, the less pleasant during your lifetime. A 13-year-old girl who spoke to the Rio Earth Summit in 1992 concluded by saying: "What you do makes me cry at night. You grown-ups say you love us but I challenge you to please make your actions look like your words."

Q Industry accounts for most of the pollution. There is little point in householders saving energy, is there?

A Carbon emissions are almost equally divided between the industrial, transport and domestic sectors. But for whom does industry make things? Largely us. Who flies and uses the roads? Largely us. It is our lifestyles that determine the amount of greenhouse gases emitted, whether directly from our dwellings or through other aspects of our lives.

This book suggests many ways to reduce your carbon footprint. They are probably best undertaken or discussed with friends or in the community, so that a low-carbon culture spreads. You can think of having a very low emission car, or no car, and do what you can to insulate your house and capture solar power. A typical family of four has one car, one house but four mouths, and its maximum potential for reducing emissions is by eating locally produced organic food. This is because conventional farming is responsible for huge emissions from transport and the manufacture of fertilisers.

The western consumerist life is a one-way process of taking resources, making them into food or things (with a lot discarded on the way) and then throwing them away as waste. As the throughput increases, the waste will gradually swamp or suffocate us, until there is no 'away' to throw them.

Landfill sites – there are 4000 in the UK – are associated with birth defects and asthma. And they are filling up. Governments are therefore turning to incineration. But incinerators, even with modern 'scrubbers', spew out microscopic particles that may cause cancer, hormone disruption and birth defects (see page 118).

The UK government has now re-branded incinerators with the euphemism Energy-from-Waste facilities and negotiates 25-year contracts with private companies that will penalise councils if they fail to produce enough waste. The counterproductive spin machine doesn't stop there. These monsters are classified – a new euphemism – as 'recycling waste', the argument being that waste becomes electricity. The government does not mention that incinerators are the most inefficient method of generating electricity, producing even more carbon dioxide per unit of energy than old-fashioned coal-fired power stations. Despite this, and despite the fact that they receive 100,000 to 200,000 objections wherever proposed, they are invariably given the go-ahead.

Our model for the new society should be found in nature, which doesn't need refuse bins to tidy it up. Nature has a circular process where growth and decay nourish an increasingly rich and complex environment.

Q Wasn't pollution much worse in Victorian times than it is now?

A The industrial revolution was built on the use of coal. Steam-powered factories replaced cottage industries, the rural population migrated to city slums, and smoke from industrial and residential chimneys filled the air, covering surfaces in grime. The rich chose to live on the western edges of towns to be up-wind from the pollution, but novels and cartoons have etched the appalling conditions suffered by most people into our collective consciousness.

Pollution takes many forms. Smoke from domestic coal fires, factory chimneys and steam trains could be seen. Emissions from cars are less obvious. Nitrates from farming invisibly pollute rivers. Dioxins from the chemical industry can't be seen but work their way up the food chain and accumulate in top predators, like us. Ozone is necessary in the stratosphere to shield us from cancer-causing ultraviolet rays, but at ground level it causes serious respiratory health problems.

Emissions of greenhouse gases are just one kind of pollution and, because they can't be seen, it is difficult to compare them with the obvious pollutions of Victorian cities. They mostly come from the burning of fossil fuels, so human-induced global warming can be traced back to the industrial revolution. This started in an isolated area but has spread throughout the world, increasing all the time. The last 50 years have been the worst for destabilising the climate.

Some hope can be taken from the dramatic action taken after the Great Smog of 1952 in London, which lasted four days and advanced the death of thousands. It was so awful and caused such an outcry that the government immediately decided to introduce the Clean Air Act, having previously said it was not possible to interfere with people's liberty in this way. It suggests that the present complacent governmental attitude to climate change could be turned round. But must we wait for a crisis to engender public outcry and frighten the government into action?

Energy Junkies

The Centre for Alternative Technology has been promoting energy conservation and the take-up of renewable energy technologies for over 30 years. Though messages about the profligate use of fossil fuels and other finite natural resources are now hitting home at local, national and international government levels, a lot of people still have questions about why we need to cut down on our energy use. It's time to stop and think about what we use and how we use it and we hope that what follows will help you to make the right choices.

Matthew Slack
Consultant

 Centre for
Alternative
Technology

Q Renewable energy can never be sufficient to power the planet, can it?

A The solar energy falling on our planet in *one hour* is greater than the energy from all the fossil fuels used by all of humankind in a *whole year*. Add in the energy of the tides, driven by the relative motion of the Earth and the Moon, and geothermal heat from deep underground, and it is clear that there is no shortage of renewable energy available to us.

The nature of renewable sources of energy is very different to the fossil fuels we currently depend on. Oil, gas and coal reserves are stored energy. Provided we can find them and extract them as fast as we want to use them, we can use as much energy as we like, when we like. We don't, however, control when the wind blows. Weather forecasting gives us some idea, but if we don't use the power of the wind when it is available, then we've missed it.

We need to change the way we think about energy to take account of these differences. We need ways of storing renewable energy in times of plenty for use in times of shortage, and more clever ways of using it so that demand is better matched to supply. Biomass fuel crops can store summer sunshine for use as winter heating fuel, but this brings with it many disadvantages (see pages 70 and 104). The electricity grid can work together with intelligent appliances to move demand for power away from peak times. But there remain some difficult areas, notably transport. There is no obvious successor to the vast quantities of easy-to-use liquid fuels that power our cars, buses and planes.

The important first step in this transition is to use less energy and to use energy more efficiently, in order to buy time to allow the supremely inventive species that is humankind to devise a safe route to our very different energy future.

Q Solar panels only work in blazing sun, how can they be useful in temperate climates?

A The term 'solar panels' covers two very different technologies. Solar thermal or solar water heating systems heat water for use in baths, showers and the kitchen sink, and are by far the most widely installed renewable energy technology in the UK. Solar electric or photovoltaic (PV) systems generate electricity. Large-scale PV systems are relatively uncommon, but individual panels are widely used to power small loads in locations without a mains supply. One of the more common uses is for lighting road signs.

We get enough sunshine in the UK to make both systems technically viable. The main barrier to greater uptake is cost.

Hot water accounts for about a quarter of the energy used in a typical UK household. A solar water heating system can provide about half of this, saving at least ten per cent of the total household energy budget. It won't do much in the winter, but in spring and autumn it will provide half the hot water, and in summer you will barely need any other form of water heating. Retrofitting a system to an existing house costs between £2000 and £5000, but if a system is installed during the construction of new houses, the cost can be as low as £1000 per house.

PV systems using current technologies are very expensive in relation to the amount of electricity they produce. To generate the 3300 kilowatt-hours (kWh) of electricity used each year in an average UK home would require a system of around 4kW peak output, costing about £24,000 at today's prices.

New, lower cost technologies are under development and when they arrive we can expect to see many more household PV systems. Until then it will remain very much cheaper to reduce the amount of electricity you use, than to install PV systems.

Q Wind power is a waste of time, inefficient and energy consuming, so why bother with it?

A The UK has better wind resources than any other European country. Nearly 2000 turbines are already installed, and they produce an average of about 30 per cent of their maximum output, double the figure managed by turbines in Germany. Larger turbines operating in the faster and more consistent winds found out at sea are expected to push this figure even higher, to 38 per cent or more.

Offshore wind farms have the additional benefit of being far less visible than their onshore counterparts. The 30 turbines of the North Hoyle wind farm off the coast of North Wales have become a tourist attraction in their own right, with boat trips taking visitors to see the 67-metre high turbines at close quarters.

Modern turbines in good windy locations generate around 60 times as much energy over their working lives as is required to manufacture, install, maintain and decommission them. This 'energy return on energy invested' is much higher than for fossil fuels. The energy equivalent of about one-fifth of the oil and gas produced worldwide is consumed by the prospecting, extraction, transport and refining processes that lie between the reserves and the end user.

With the exception of gas-fired power stations, wind power is the cheapest form of new electricity generating capacity. Once turbines are installed there are no fuel costs and decommissioning is cheap and virtually instant. The land surrounding the turbines can continue in its original use.

Recent studies into the variability of wind energy have concluded that wind can make a reliable contribution to our energy needs. It is less good at providing power on demand. With wind power making up less than four per cent of our current electricity supply, this does not yet present any difficulties. When we reach higher proportions of wind power, we will require a cleverer approach to match supply with demand.

Q Energy saving devices such as light bulbs are always ugly, consume energy in manufacture and are more expensive. Why change?

A Since the European Union announced the phase-out of conventional incandescent light bulbs by 2009, major retailers have slashed the prices of low energy 'compact fluorescent' bulbs (CFLs). They often now cost less than £2 each. With their 80 per cent lower running costs and up to ten times longer life expectancy, overall CFLs not only save energy, they save you a considerable amount of money.

Lighting often accounts for 20 per cent of the household electricity bill and replacing all lights with low energy versions can save 80 per cent of this. That's a 16 per cent reduction in the electricity bill for an expenditure of just a few pounds.

The range of both size and shape has broadened considerably in recent years, so that a suitable bulb is available for most situations. It is no longer necessary to have an ugly, tubular affair protruding from the bottom of your lampshade. Look out for the more compact coiled versions, and the ones enclosed within a conventional bulb-shaped outer casing.

For spotlighting applications, have a look at the other developing low-energy lighting technology – 'light emitting diode' lamps, or LEDs. Early ones were of limited power, but they are developing fast. LEDs are as efficient as CFLs but provide a more directional light.

Some energy is consumed in making all kinds of bulb. The longer life of CFLs compensates for their higher 'embodied energy'. In future, the electronic controls built into each bulb will move to the light fitting, further reducing this aspect of their environmental impact.

Other energy-saving electrical appliances look no different to their conventional counterparts. The lowest energy refrigerators and freezers now have an energy label rating of 'A++', which means they use at least 45 per cent less electricity than even an 'A' rated version. Compared to a typical ten-year-old appliance the saving will be 75 per cent or more.

Q Nuclear power is the answer as it creates no carbon emissions. Is it a clean, safe solution?

A Some carbon emissions are involved in the building and decommissioning of nuclear power stations, and in the complicated process of manufacturing nuclear fuel from uranium ore. It must therefore be seen as a low, rather than zero, carbon energy source.

Whereas in normal operation nuclear power stations may seem to be safe, the consequences of a major nuclear accident are very severe, and there have been several. Modern designs will undoubtedly be safer than the existing reactors, but after the 2001 attacks on the World Trade Centre it is no longer unthinkable that a nuclear power station could become the target for a terrorist attack.

The radioactive wastes produced by nuclear reactors are highly dangerous and will need to be securely stored for thousands of years. Do we really want to burden future generations with this?

Other difficulties facing nuclear power as part of our response to climate change are the very long delay from deciding to build to generating power, and the sheer cost. If we start tomorrow, there is unlikely to be any new capacity online before 2020, even though the go-ahead for a nuclear power programme has now been given by the UK Prime Minister. Climate change is more urgent than that. On cost, we are already facing a £73 billion bill to decommission the existing stations. Private companies are unwilling to invest in new nuclear power stations without the promise of substantial, long-term government financial support. This would take much needed funds away from the developing renewable technologies.

We should not overestimate the contribution of the existing fleet of nuclear power stations. While they produce a sizeable 18 per cent of our electricity, they contribute nothing to the oil we need for transport and the gas most of us use for heating. Their share of our total energy consumption is just four per cent.

The most we can expect from new nuclear power stations is eventually to match the current capacity. They do not offer a quick, cheap route to large amounts of low carbon energy.

Q What about nuclear fusion?

A Our existing nuclear power stations use the process of fission – high molecular weight atoms of radioactive elements such as uranium and plutonium split by the natural process of radioactive decay into atoms of lower molecular weight, releasing energy in the process. Fusion is very different. Atoms of very low molecular weight – hydrogen – are joined together to form the heavier atoms of helium. Again, energy is released as heat during this reaction.

This all sounds very simple, and seems to overcome the dangers of radioactivity associated with nuclear fission. There are however enormous technical difficulties to be overcome. The fusion reaction is what makes the Sun work, and it occurs at staggeringly high temperatures – literally millions of degrees Celsius. No known material can tolerate these conditions, so that the fusion reaction has to take place supported by a strong magnetic field, away from the walls of the physical container.

The difficulties in making fusion work are amply demonstrated by the fact that research has been going on for decades, but the most successful tests so far have only managed to maintain the reaction for a matter of seconds, and in doing so have always consumed more energy than is produced.

It will be at least 30 years, and probably more like 50, before any commercial fusion reactor contributes to the national grid, which means it is not going to be ready in time to help with climate change. Assuming that we manage to get through to 2050 with a working planet, fusion may well be part of the energy supply in the second part of this century. In the meantime, it must not starve technologies with more immediate benefits of the research and development funding that they require.

Q Drying my clothes and washing my dishes by machine is the only way I can fit these chores in to my day. Do I have to give up these machines?

A There's no need to worry too much about using your dishwasher. Used properly, it will use less water than washing up by hand, and in most cases less energy too. Select the lowest temperature setting that will do the job, and always run the machine full. Of course, if you rinse everything beforehand and afterwards the water saving is lost; and if you do this in hot water you will use more energy too.

Tumble driers are another matter; they work by evaporating the water out of your clothes – the amount of energy needed to do this is fixed by the laws of physics. It takes about five times as much energy to evaporate a litre of water as to boil it in a kettle – no wonder driers are such energy hogs. That's why it is so worthwhile to dry clothes naturally whenever possible. But if you really can't, there are two ways of reducing the carbon.

First of all, reduce the amount of water that needs to be removed. The higher the spin speed of your washing machine, the lower the water content of the clothes when they go into the drier. A washing machine with an energy rating of 'A' for spin performance will leave your washing one-third drier than an average 'B' rated machine.

Then, provide the heat needed by the machine in a more carbon-efficient way. Burning gas directly provides heat with less than half the carbon emissions of electrical heating, and although gas driers require a registered installer, they are no more expensive than conventional electric models. Alternatively, advanced electrical driers incorporating heat pump technology reduce electricity consumption by half, saving enough money over their lifetimes to offset the higher purchase price.

Finally, don't forget that one drying cycle is equivalent in carbon emissions to driving an average car six miles!

Q I need to keep warm at home by turning up the temperature. What's the alternative?

A Most people are comfortable at a room temperature of 18°C. Heating to higher temperatures adds between five and ten per cent to your bill for each degree you turn up the thermostat.

If your house does not feel warm and comfortable, then check the basics before adjusting the thermostat. Your loft should have at least 270 millimetres of insulation. If you have cavity walls then in 90 per cent of cases these can be insulated. This alone can save a quarter of your heating bill. Check for draughts, especially around windows, doors and through bare, boarded floors. If you don't have double-glazing then heavy, lined curtains will make a difference. If you have a hot water cylinder, make sure it has a good, thick insulating jacket in place. There's no harm in adding a second layer.

Your boiler consumes around 85 per cent of the energy used in your home to provide central heating and hot water. A modern condensing boiler will use about one-third less fuel than the average ten-year-old boiler. If you combine the basic insulation measures with a new boiler, often the heating costs can be halved.

The solid walls of older houses are more difficult to insulate, but doing so can make them as good as a modern house. Internal insulation takes up a small amount of space indoors and requires redecoration afterwards. External insulation changes the external appearance and may need planning permission. Many useful guidelines can be found on the websites covering energy saving in the resources section of this book (see page 155).

Q **I leave my TV and other appliances on standby as I believe it uses less energy than switching on and off. Why should I turn it off?**

A It is surprising how much power appliances waste in standby mode. Taking the UK as a whole, in one year standby accounts for the entire output of an entire power station. The cost to an average household is around £38 per year.

A traditional, boxy television set uses around 65 watts (W) when in use. On standby this drops to around 7W. If on average it is watched for four hours a day, then one-third of its power consumption is wasted during the other 20 hours. There are worse offenders. How many power tool battery chargers lurk in sheds and garages? Used infrequently, these often consume a steady 12W year round. Every watt of unnecessary standby power consumption costs about £1 per year.

Many electrical appliances continue using power even when you think you have switched them off. For example, when you shut down your computer, it appears to turn itself off. If you then hit the monitor off switch as well, you can be forgiven for thinking that those devices are no longer consuming power. In most cases this is not so. This 'phantom' power consumption for the PC and its monitor can be high as 30W. For a home PC, used for two hours per day, the wasted energy can be two-thirds or more of the total.

The simple way to avoid standby is to switch everything off at the wall when not in use. If the sockets are hard to get at then use an extension lead to bring them within reach. You can get versions with individually switched sockets, allowing only the devices you are using to be switched on.

For the ultimate standby eradication programme, get a power meter which will allow you to measure the performance of every appliance in your home.

Q Is it true that vegetarians use less energy?

A The food supply chain in the UK accounts for around a quarter of all our greenhouse gas emissions. Only a small proportion of this is directly attributable to agriculture. By far the largest share is transport and around half of this is the final journey from supermarket to home.

Changing your shopping habits is potentially as important as changing your diet. Buying local food will significantly save energy in terms of transport.

Agriculture is responsible for major shares of our nitrous oxide (N_2O) and methane (CH_4) emissions. These are respectively 300 and 20 times more potent greenhouse gases than carbon dioxide (CO_2). All cultivation and use of artificial nitrogen fertilisers releases nitrous oxide. Livestock, in particular ruminants such as cattle and sheep, are responsible for significant levels of methane production (see page 82).

A vegetarian diet does not necessarily make much difference. Protein intake is often obtained from increased consumption of eggs, milk and cheese, and these products rely on exactly the same industry as the meat they replace.

A vegan diet can make a substantial difference, not only to energy use but to the other pressing food-related issue: land use. It has been estimated that a vegan can live on just one-fifth of the land required by a meat eater. By eliminating the need to grow vast quantities of animal feed, land can be released either to feed a larger population or to grow increasing quantities of energy crops (see page 104).

It is estimated that for every calorie of food you eat, ten calories worth of oil has been consumed in the planting, fertilising, harvesting, processing and transporting of it. You can make worthwhile reductions in that by choosing locally produced food, eating less meat and dairy products and avoiding highly processed foodstuffs.

Finally, always compost food waste rather than putting it in the bin. Decomposing organic matter in landfill sites is another major source of methane emissions (see page 126).

Q If air miles are an issue, how can we support Fairtrade?

A Aviation is completely dependent on dwindling oil reserves, and is also the fastest growing source of carbon dioxide emissions worldwide. If we want to retain the benefits of air travel and transport then we need to use it sparingly, and only when there is no other way.

Only a minority of Fairtrade produce arrives by air. Tea, coffee, cocoa for chocolate, sugar and bananas travel in bulk by sea, the mode of transport with the lowest carbon dioxide emissions (see page 91).

The business case for air-freighted food, whether Fairtrade or not, relies on the premium prices that rich people in developed countries are prepared to pay for out-of-season luxuries. If aviation fuel were taxed, and the environmental cost taken into account instead of just the financial cost, the case would look rather shaky.

Climate change is the most pressing environmental problem facing the world today. But with two-thirds of the world still living in poverty it is understandable that it is not top of the agenda in developing countries. Developing countries need to build a long-term future. Business models that rely on the most environmentally damaging form of transport may bring short-term financial gains, but they are inherently unsustainable.

Fairtrade is a great idea, but it is only needed because the way in which world trade is organised favours the developed countries and penalises the less developed. All trade should be fair, but it isn't. Developing countries see very little of the profits that accrue from the foodstuffs they export, because most of the value is added later in another country. Exporting chocolate instead of raw cocoa dramatically increases the proportion of the profits that stay in the country of origin (see page 96).

Travel
Turbulence

As I write there are 28,000 bags piled up at Heathrow. The baggage system at Heathrow's new Terminal 5 has gone wrong and has caused chaos. It is hard not to sympathise with the poor folk who have been stranded or who have gone abroad without any knickers.

But there is a touch of 'I told you so' about it, too. Might we deserve a bit of chaos for insisting on flying all over the world with scant regard for the consequences? Those consequences can be horrendous, and it is now becoming a moral choice to fly so much. If that sounds tough and judgemental, read the arguments below.

Alastair Sawday
Publisher

$awday's Special Places to Stay

Q How are people going to see the world and gain life experience, without low-cost travel?

A Marco Polo got to China, and Alexander the Great got a whole army as far as India, long before air travel. (Admittedly, Alexander didn't take his family with him.) Before low-cost flights there were hordes of eager Europeans and Americans cruising around the world; it was hard to get away from them in some places. Before them, the Victorians swamped parts of the world – such as the south coast of France, where hotels were called Hôtel d'Angleterre and seafronts were called Promenade des Anglais – as in Nice. It wasn't just Europe that the Victorians went to; they colonized much of the world, sometimes in large numbers.

An enterprising young Englishman called Ed Gillespie took off around the world in 2007 to remind us all that travel need not be expensive or flight-bound. He went by train and ship, never flying, and had a ball. Once upon a time hitch-hiking was considered a viable means of transport, and it was. You could get about for nothing more than an occasional contribution to petrol costs. It is a shame that the tradition has almost died, but those who still do it are reminders that you don't need to be rich to travel. Indeed there is another young Englishman who is attempting to walk to India without a penny on him. I'm sure he will eventually succeed.

Life experience is gained wherever we go. The richer our imaginations, the richer our experiences. It isn't just travelling, of course, that enriches us, and often travelling doesn't enrich us at all. A rowdy weekend in Ibiza probably gets us nowhere, neither does a package holiday to Sharm el Sheikh. But an hour under falling autumn leaves in the Chilterns may well do more for us. It is a triumph of modern advertising that we are now convinced that we have to jet away to another country in order to gather life experience.

Q Flying only contributes three per cent of the UK's carbon emissions, so what's the problem? Surely it would be banned if it were that bad?

A If it really were three per cent then there would be slightly less concern about it. But the figure is closer to 12 per cent if you take everything into account, such as the damage done by water-vapour and carbon at the very high levels where aircraft fly. Equally to the point, air travel is growing very fast so if we don't curb it there will be a far greater problem to tackle later on.

If carbon pollution is a 'bad thing' in itself why argue for more of it – as the aircraft and airline industries do? Most of us would agree that litter-dropping is to be discouraged, and even banned. Who would argue that it be allowed to increase because we only do a little of it? The same is true of alcohol consumption. Who would argue for an increase other than those who make the stuff?

The difficulty for the environmental case with aviation is that so many of us are now 'enjoying' the benefits of flying that we have become the 'vested interest'. We see ourselves as flyers rather than as citizens, so have lost sight of the bigger picture. But that is a common problem and applies to car-ownership as well.

Of course there are economic arguments in favour of boosting aviation. But the UK is perfectly capable of being an economic powerhouse without subjecting itself to (apparently limitless) airport expansion. Can we not invest in, say, renewable energy technology rather than in airports? Might it not be a better and more visionary path? What happens when the oil price soars even higher, as it certainly will? Will we be looking very silly with our thousands of hectares covered by largely empty tarmac that we cannot afford to use? Countries are littered with abandoned industrial sites: coal mines, shipyards, steel works. Abandoned airports are inevitable. Why build more?

There is a powerful moral case, too, for looking critically at a mode of transport that, in the case of one person's return flight from Europe to the Caribbean, for example, generates as much carbon dioxide (CO_2) as the average family, worldwide, generates in a whole year. Again, taking not the world average but the carbon emissions of an average African, that return flight does as much damage as one Tanzanian does in 40 years.

Returning to the original question: aviation emissions doubled between 1990 and 2000. The government estimates that there will be another doubling by 2030. Even the government accepts that carbon emissions must be curbed, dramatically.

Would it not already be banned if it were 'that bad'? For an answer to that you need only look at cigarette smoking! It takes time for politics to catch up with the science, especially if there are people who don't want them to. Don't imagine that the aircraft industry and airlines will take the arguments lying down.

Q If 300 people who would have taken one plane could instead drive individually to the same destination, which would be the greener way to travel?

A A large American saloon car taking part in a rodeo show in the 1960s entered the ring under its own power and disgorged 42 people – and a donkey! The audience was impressed, and had witnessed an unusually green form of transport.

If you stuff a lot of people into one vehicle you are equipping it to compete well in the green arena – unless it is a plane. A train is not always greener than a car, especially if it is largely empty while the car is full. But the question is more complex, for everything depends on where this full plane would have flown and where the cars are going – and what sort of car it is. For example, a little diesel car running on recycled cooking oil is probably more efficient than a half-empty bus.

To calculate exactly which form of transport would be greener for any given journey could be immensely difficult. So it is best to fall back on some basic assumptions and guidelines. Thus: biking is even more efficient than walking, for a turn of the pedal can take you several metres while the same effort in walking would take you just one metre. The energy used to make the bike is pretty minimal and the bike can last for years, and even decades. (My spare bike is a five-gear Sun, about 40 years old and still going strong. It never breaks down.) Next best could be a motorbike with a smallish engine, driven carefully, and then perhaps a very efficient small car or a train. Then comes a bus and lastly comes a car. Even further behind, hardly worth considering, is a plane. But then, it all depends...

To show you how complex this question is, here are some figures: a Toyota Prius, the first mass-market hybrid car, emits 104 grams (g) of carbon every kilometre. A Toyota Yaris and a tiny Smart car emit 127g, a VW Golf 2.01 emits 194g, a Landrover Discovery 275g and a Lamborghini 500g! So, if you drive a Golf from London to Glasgow you will emit 128 kilograms (kg). If you fly you will cause emissions of 336kg and if you take a train that will be 72kg. (All these figures are approximate.)

Now, back to the original question. 300 people driving individually will emit about 38 tonnes of carbon (300 × 128kg). Flying together they will emit over 100 tonnes (300 × 336kg). Convinced?

Q The flights are going anyway, so they may as well be full.

A The flights are going because they are full! They are not going for the sake of it. The flights will only go if there are enough passengers on them regularly to make it worth flying. Airlines are immensely cost-conscious and watch their seat occupancy like hawks, especially the low-cost airlines. You will find that they will soon stop flying into a small town in Europe if it isn't worthwhile.

Q There is no alternative to flying, especially for cross-continental travel. You can't expect people to go back to boats.

A Two journalists raced from central London to central Paris. The one who went by train (using Eurostar) got there before the one who flew, and the cost was roughly the same.

Until the 1990s most people who went to France and onwards to other continental countries went by ferry. The ferry companies operated from lots of southern ports and competed vigorously with each other. Then came the Channel Tunnel and the competition got hotter. The ferry companies slashed their prices and became even cheaper and more comfortable. People would even pop across to France for the weekend, though generally we took longer foreign holidays. Once in France, of course, it was considered easy to drive long distances. Or you could pop the car on the train – or travel by train without it, anywhere and easily.

Then along came the low-cost airlines. Within a few years we have got used to nipping across the Channel by air, at a frequency unthinkable a few years ago – so much so that airports are now becoming crowded and unpleasant. If you have ever been badly delayed, or diverted, or held up by security checks, you will know what I mean. In fact, given the need to get to airports earlier and earlier before a flight, the total travel time for a lot of continental journeys can be greater by air than by rail and even by ferry in some cases. (Try flying to Normandy.)

It is true that to travel right across Europe to, say, Hungary is bound to be quicker by air than by train and ferry. But speed isn't everything, is it? There is a now growing movement towards 'slow travel', getting from place to place at a pace that creates no stress and is actually enjoyable. Many people are taking the train to Italy, say, rather than the plane; and counting their blessings. The process of travelling can be half the enjoyment. A comfortable ferry journey followed by a train ride is something to look forward to. There is no reason why we should be frightened of switching from flying. It might even be advisable to start practising – for soon flying will cost a great deal more than it does now. Oil is now over $100 a barrel, unthinkable a while ago. Some predict that it will double in price within a year or so.

Q I only get a couple of weeks holiday a year, I don't have time for green and therefore slow travel options.

A If time is short, flying doesn't make it any longer. Do you really not have time? Or are you feeling pressured because all your friends are rushing off to distant places by plane? Think back to the myriad ways in which civilised people have taken and enjoyed holidays for hundreds of years. Our modern devotion to air travel is skewing our judgement. If Richard Branson ever establishes space travel as 'normal' for the very rich, they will feel left out if they are not doing it.

Perhaps we can reclaim our judgement and really ponder the meaning of holiday and relaxation. It is, admittedly, hard to do so when we are assaulted by fashion and by advertisements from budget airlines. But if you have only a few weeks of holiday, here are some suggestions:

- How about resolving to get the most out of Britain? Other people spend fortunes coming here for their holidays, even if it does rain. Youth hostelling is cheap and can take you to glorious countryside. If you need more comfort, Britain's hotels are now pretty impressive and can be found in the remotest places. We have it all: heritage, architecture, beauty, culture, wild places. Roger Deakin, a British writer, spent his time swimming in the thousands of lakes and rivers of Britain. He was puzzled by his friends' determination to go further away.

- Divide the time up into small chunks and add them to your weekends, thus maximising the impact. A good, long weekend away can have the restorative effect of a holiday. In fact it is a holiday! Five days of holiday can be used to create two great weekends, leaving work at lunch-time on Thursday and returning on Tuesday. You get two holidays for the time-price of one. You could even make those weekends five-nighters if you play your cards right.

- For those weekends you can take the train anywhere in western Europe: Paris, Brussels, Amsterdam, Bruges, Strasbourg, Rouen. The list is long. You can even go as far as Berlin. Or leave London on Thursday afternoon and get to Milan on Friday morning for a three-night break in Italy, returning by Tuesday morning in time for work.

- If you live near the south coast it is easy to pop across to France by ferry. Within hours you are deep in France – a tonic.

Even more radical: why not stay at home and ignore all the jobs that hang over you? Throw a blanket over the piles of papers you need to sort through and go play with your friends: go to the cinema several times, eat out, walk around your own city, bike into the countryside, spend time in the pub. Or even just read a few books.

Q I need my 4×4 to get the kids to school safely, with all the traffic on the roads.

A Do you? And if your children appear to be safer in your 4×4, have you wondered how much less safe everybody else is once you start driving it? It is this sort of thinking that has led many American parents to buy Humvees – vehicles more like tanks than ordinary cars. In my madder moments I can almost imagine a world where parents actually do drive tanks to school, flattening cars and even people – and ensuring the safety of their occupants.

George Monbiot, a British writer, reported that because big 4×4s are higher and heavier, the occupants of a vehicle hit by one are 27 times more likely to be killed (according to the US Insurance Institute for Highway Safety) than the occupants of a vehicle hit by a normal car. For the same reasons, they kill between two and three times as many of the pedestrians and cyclists they hit than do smaller cars.

By the way, 4×4s are big polluters. Driving one for a year (average driving) wastes as much energy as leaving your fridge door open for seven years! They pump out about 2.5 times as much carbon as a small car, take a huge amount of energy to make, and – of course – use up more than their fair share of petrol and oil.

These giant cars are also more likely to roll over in an accident than lower cars – which seems obvious when you look at them, just as a double-decker bus is more likely to roll than a single decker. So if you value the lives of other people's kids, and your own too, then you might be better off teaching your kids to cycle safely to school. That way they learn a bit of independence, get some exercise and avoid being spoiled. Mum and Dad can sit a bit longer over breakfast and avoid the strain of jostling for space on the road.

Q Cycling is too slow and dangerous.

A A reasonably fit cyclist can bowl along for hours at an average speed of 16 kilometres per hour (km/h). His or her average speed in London, say, is probably 10–13km/h. The average motorist in London bowls along at an average speed of 6km/h. And that is without counting the time spent looking for a parking space.

Outside towns and cities the car has the speed advantage, of course. But even then it is not as obvious as it may appear. For example, I cycle 4km to work every day and take 17 minutes – a bit slow because there are two big hills. My colleagues who drive take about ten minutes, not a huge advantage. Meanwhile, I have the added advantages of exercise, low cost and a delightful ride through a deer park and woodland.

The vexed question of safety is a harder one to answer. Clearly cyclists are more vulnerable than motorists and it is hard to find meaningful statistics because the accidents that cyclists have are very different from those that drivers have. I would agree that urban cyclists are more likely to have accidents than urban motorists. But a very careful urban cyclist is as safe as a motorist; the trouble is that many of them are a bit reckless. I know, for I am sometimes.

Interestingly, there are statistics from Denmark which suggest that cycling is safer than driving, because cyclists live longer than inactive drivers. So, putting this in a broader health context, we should be encouraging cycling.

It is generally believed that cyclists live ten years longer than inactive non-cyclists. That strikes me as 'safe'.

Q Now everyone is growing biofuels to fuel transport, isn't the world running out of wheat?

A Yes, and this may well lead to international catastrophe, and even wars. It has already led to hunger, and rising costs for the very poor.

The sadness is that some of the motive for switching to biofuels was genuinely 'green'. But there has been a massive misunderstanding of what biofuels involve and, indeed, what they are.

Biofuels are fuels made from biological material – such as sugar cane, corn, olive oil, wheat, wood and so on. For decades the Brazilians have grown sugar cane to make ethanol, a biofuel that has served them brilliantly and saved them a lot of oil. But Brazil has a lot of space and a lot of sugarcane and can afford to devote the land to growing it.

If land that has been used for growing crops to feed us is then used to grow crops to power our cars and lorries, where will the food come from? This question becomes urgent when food is in scarce supply. Once upon a time there was more grain grown worldwide than was needed, so we stored a lot of it to tide us through hard times. Now, for the first time in recorded history, the world's demand for grain has outstripped the supply. That is a terrifying fact. Someone is going to go hungry, and you can bet your bottom dollar that it will not be the rich. It is not extreme to suggest that for every one per cent rise in the cost of food, vast numbers will starve.

So, growing cereal and food crops for fuel is, generally, a form of latter-day madness that may set fuel-hungry nations against hungry ones. Environmentalists have never argued in favour of this.

However, biofuel does have a place; environmentalists know this and have long argued for its intelligent and appropriate use. For example, where there is surplus timber why not use it to power heating units, such as boilers and fires? It is renewable and takes the place of fossil fuels, which once burned are lost forever. Other biofuels include wood-pellets made from sawdust that might otherwise go to waste. Perhaps the best of all, though there are limited supplies, is recycled cooking oil. It has been used and then serves a second purpose; its disposal would otherwise cause problems; it is easy to clean and convert to biofuel and actually does diesel engines some good. Lastly, it can be mixed with diesel in any proportions.

The question supposes that environmentalists have got it all wrong. The truth is that others have jumped on to a passing bandwagon without seeing which way it was heading.

Q Tourism is essential for developing countries. Why should we deprive them of the affluence we enjoy?

A Any activity seems to become essential once we do a lot of it. The more fish I eat, the more I will think it is essential to me. The more tourists that we send to The Gambia the more The Gambia will become dependent on them. So the argument is slightly self-promoting. It is important to cut through this vicious circle and have a serious look at what tourism actually achieves in developing countries.

- Tourism takes people away from traditional work, such as fishing. In coastal fishing communities it is commonly the case that fishermen become waiters, that there is no fish for the community so they have to buy it in from elsewhere. For this the community needs cash, which they can only get by working in the cash economy – so they, too, become waiters. Thus a community loses its independence, its autonomy and its freedom.

- Most investment in tourist development is foreign, and the profits, of course, go back to where the investment came from. Little trickles through to the local people, who watch as their beaches are taken over, their land is turned into golf courses, their water table is lowered by hotel use and their children become low-paid waiters or mountain porters.
- If the tourist market moves on, the locals are left with the damage and no source of income. Spain is beginning to face this as the package tours go elsewhere and they are left with thousands of hotel bedrooms to fill.
- The first priority of any proud community must be to retain that pride and its ability to look after itself. Tourism undermines these things. It treats the local people as servants, objects of curiosity, disposable tools for generating profit. There are few developing countries with the institutions to resist tourism and make it work for the good of the inhabitants rather than for its investors. Even Thailand is overwhelmed by the sickening effect of the most cynical form of tourism – child prostitution. And adult prostitution too.

When talking about affluence, we should ask ourselves why we, hugely affluent and thus able to fly away on holiday, want to go to developing countries. Is it because they have something that we haven't got? Like peace, a slow pace, beautiful scenery, fresh food, colour and community? Will the affluence that we use to escape from our affluent societies on holiday enable the locals to have more of those things? Or would it change them forever? If so, I suggest that they will be puppets at the end of our strings – and look how our history tells us we would use the pull.

A holiday in Ladakh, part of Kashmir, can show you that our affluence indeed enables us to get there and learn from the Ladakhis. But we soon realise that it is their very lack of affluence, their utter contentment with the lives they lead and have led for centuries, that impresses us. They have what they need; we are looking for more and more – and more.

Q Why should I holiday in the UK? The weather is terrible.

A Fair enough, especially in 2007! But don't forget that while we were getting depressed under the grey skies of August, those who had fled to Greece were sweltering under temperatures of 43°C. And so it was in many Mediterranean countries. It was unbearably hot all over Europe.

As climate change takes hold and alters our climate in unpredictable ways, it is clear that weather all over the world is going to become less and less certain. It is becoming harder and harder to be sure that going abroad for our holidays is going to guarantee better weather. So why not stay here?

Holidaying in the UK has never been more interesting or exciting. Hotels and 'bed and breakfasts' are a world away from the grim places of our folk-tales. They are often as good as anywhere you will find abroad. Our coast is stunning, our beaches as good as any. (My own world favourite is in Cornwall.) We have great houses to explore, the world's most seductively lovely countryside – partly because there is so much rain. Even our food is now good, sometimes wonderful. And where else can you find such great beer?

But the best reason for staying in the UK is that you don't have to go far, hang around in airports or spend all that money on fares. The canny among us stay in the UK, spend the saved money on good food, and learn what a beautiful place this is.

Funny Farm

The Soil Association is the UK's leading environmental charity campaigning for people and planet friendly, sustainable organic food and farming.

Concerned about the impact of intensive farming practices on the environment, food quality and health, the Soil Association challenges the dominance of industrial agriculture, connected as it is with degradation of the countryside, increases in diet-related illness, mistreatment of animals and the betrayal of public trust in food. We offer practical solutions to everyone involved in the food chain – farmers, food processors, retailers and consumers.

We certify over 70 per cent of the organic food, drink, textiles, health and beauty products sold in the UK, guaranteeing these have been produced to the highest standards.

The Soil Association is a membership organisation. For more information visit www.soilassociation.org. Our consumer website www.whyorganic.org holds a directory to source organic food locally.

Robin Maynard
Director of Communications

 **Soil Association**

Q Battery farming standards are strict and protect us and the animals. Isn't this the only way to provide cheap food for the population?

A The two farm animals that have suffered most under the development of industrial farming methods are certainly the chicken and the pig. Confined to window-less sheds in their tens of thousands, in some cases hundreds of thousands, these animals whose ancestors were jungle fowl and wild boar have had their natural behaviour more modified and restricted than any other domesticated livestock. True, factory-farming succeeds in raising millions of pigs and poultry to the point of slaughter, and producing vast quantities of 'cheap meat' unimaginable to pre-war generations. But in doing so it compromises animal welfare and causes potential risks to human health.

Take the battery-cage, typically holding four to five chickens. Each bird's space is smaller than an A4 sheet of paper. Over 200 million laying hens are kept like this across Europe. Through frustration, boredom and close proximity to each other, hens often peck their cage-mates. Pecking can lead to cannibalism. In an attempt to stop this, hens are often de-beaked – one-third of the bird's beak is removed using a red-hot blade. Similarly, bored pigs bite each other's tails. This stress response is 'managed' in intensive systems by simply cutting off their tails. The animals are thus physically altered (mutilated) to suit the farming system, rather than the system being adjusted to best meet their natural behaviour patterns. Osteoporosis and fractured bones are common in caged hens because the high rate of egg production depletes the hens' calcium reserves. The wire floor of a battery cage causes the hen's claws to become twisted around the mesh. Damaged legs and feet further reduce the hen's limited opportunity for exercise, sometimes preventing her even from reaching the food and water supply.

The good news is that battery-cages as they are now will be prohibited across Europe from 2012. But the replacement, EU-approved designs are far from free-range. Often referred to as 'enriched cages', these modified battery cages provide more floor space per hen, but this still amounts to less than the area of an A4 sheet. Leading animal welfare group Compassion in World Farming is not convinced, "Although the provision of a nest-box for the hen to lay her eggs in is a real improvement on the barren battery cage, the provision of a small litter area and very low perches do not make a sufficient improvement in hen welfare... Enriched cages are still crowded and severely restrict the hens' movement and natural behaviour". 'Enriched' cages may hold fewer than ten or up to 60 or more hens, depending on their size and design. However, research shows that any space allowance less than 5000cm^2 per hen (nearly ten times that provided), restricts natural behaviour to some extent.

Q Do the pharmaceutical drugs given to animals lead to superbugs in people?

A Keeping very large numbers of stressed animals crammed together in close proximity creates perfect conditions for disease. To combat this, antibiotics are routinely used to keep intensively-reared animals alive. Until recently, they were also legally added to chicken and pig feed as 'growth-promoters'; by knocking out 'bad' bacteria in the gut, they improved the efficiency of the animals' conversion of feed to meat.

Over 30 years ago, the Swann Committee identified factory farming's role in creating antibiotic-resistant super-bugs. Set up following an epidemic of food poisoning from salmonella resistant to several antibiotics used in human medicine, the inquiry headed by Lord Swann urged complete separation of drugs used in human medicine from those used to treat livestock – advice that was ignored by politicians under pressure from the intensive livestock lobby.

Since then multi-drug resistance has increased at an alarming rate, with some salmonella strains now expressing 95 per cent multi-resistance compared to just five per cent 20 years ago. Three decades on from Swann, the use of antibiotics as growth-promoters in livestock production has finally been banned across Europe, but that hasn't ended widespread, routine recourse to antibiotics, nor their knock-on impacts on human health. In 2007, the Soil Association revealed the discovery of a new strain of MRSA found in intensively farmed pigs, chickens and other livestock in Europe. Described by one scientist as 'a new monster', farm animal MRSA has already transferred to farmers and their families in The Netherlands. As many as 50 per cent of Dutch pig farmers have been found to be carrying it. If you're a Dutch pig farmer, or from a pig-farming family and need to go into hospital, expect to be immediately isolated and swabbed to detect whether you are likely to be carrying the new strain. It has not yet been confirmed in UK livestock, but then our government isn't really looking for it.

There is plenty of evidence for the lowering of immune systems through the stress of close, unnatural confinement and for the 'amplifier effect', on large numbers of animals or birds kept in close proximity, on the virulence of a disease. Keeping livestock closely packed indoors has been described as providing 'a Petri dish for diseases to circulate and infect other birds'.

Commenting on outbreaks of Newcastle disease (a poultry disease) in Australia, a government leaflet was unequivocal in pointing to large-scale commercial units, rather than backyard or low-density outdoor keepers, as the cause of increased virulence. It stated that a very mild form of Newcastle disease virus was present in all States, but providing that the strain did not mutate into something virulent, it posed no threat to birds. Outbreaks on the mainland between 1998 and 2002 were attributed to a mutation of the endemic mild strain into a virulent strain of the virus. It concluded that the available evidence indicated that, for such a mutation to occur, it needed a large number of birds in a small area – the conditions found in intensive farming.

Factory farming is not just breeding super-bugs – it is also under suspicion for creating new, more dangerous viruses, such as the latest, highly pathogenic form of bird flu – H5N1 – threatening the UK poultry industry and wild bird populations. The campaign organisation GRAIN (an international non-governmental organisation that promotes the sustainable management and use of agricultural biodiversity) has tracked the development and spread of H5N1 back to the vast and rapidly expanding intensive poultry units of the Far East. The cramped, humid conditions of an intensive broiler unit with tens of thousands of chickens crammed together, provides the perfect evolutionary journey for a virus from a mild to more deadly form. According to GRAIN, this new strain of bird flu has travelled west along the roads and railway networks used by the global poultry trade rather than via the migration routes of wild birds, which until now have offered the international industrial poultry business a suspiciously convenient scapegoat.

Q Is organic meat better for us, or just over-priced?

A An endless cycle of diseases seem to besiege UK agriculture, from BSE (mad cow disease) to Foot and Mouth, to the more recent concerns over bird flu, bovine tuberculosis (TB) and Bluetongue in cattle. Animals reared organically can contract diseases, just like their intensively reared, non-organic cousins. But there is evidence that they can be more resistant and are often better able to recover. BSE is different: it is generally accepted that the practice of 'recycling' animal protein back to cattle through processed cattle feed was the key cause of BSE. There has not been a confirmed case of BSE in any organic livestock raised on a fully converted organic farm, most probably because feeding animal protein to cattle and sheep – vegetarian, ruminant animals not designed to eat meat – is not permitted under Soil Association standards.

The idea that healthy, well cared for animals are more resistant to disease is not a new one. Sir Albert Howard, working as a colonial agriculturist in India at the turn of the 20th century, developed his theory of 'positive health' influenced by peasant farming practices. He observed that the ability of plants to resist pests and diseases was directly linked to the health and fertility of the soil in which they grew. Similarly the vitality of livestock was linked to the quality of the food they ate. To test this, Howard deliberately exposed his cattle (fed with well-composted crops) to neighbouring animals infected with Foot and Mouth disease and scientifically recorded the outcome. The results are compelling: between 1910 and 1936, at three different locations, a total of 58 cattle reared on organically managed land were allowed to come into contact with infected animals. In only one year of the 26 did a few mild cases of Foot and Mouth occur – and this was a particularly hot year, during which the animals were worked hard and their forage was poor due to drought. Yet even these infected animals recovered well, going on to win prizes at agricultural shows.

Bovine TB is the most controversial, current disease afflicting UK farming. If a dairy cow is found to be infected (known as a 'reactor') the farmer's whole herd will be restricted, with potentially devastating impacts on the farm business. Bovine TB is spreading throughout the dairy areas of the UK; testing and culling cattle and compensating farmers is costing a fortune – around £90 million annually. While the movement of cattle is the confirmed key cause of the spread, wildlife, in particular badgers, is also held responsible as a 'reservoir' for the disease. Frustrated at lack of any clear action from government, many farmers are demanding a mass cull of badgers to reduce one suspected route of infection. But this polarised debate is diverting attention from what may well be the underlying causes, the same ones known to exacerbate the incidence of TB in humans – poor diet, cramped, damp living conditions, stress and a generally depressed immune system.

The modern dairy cow has been bred to be a 'milk-machine' producing 10,000 litres of milk annually, compared to the average of 3500 litres per cow 30 years ago. With the metabolic rate of a Tour de France cyclist pedalling uphill, many dairy cows are culled exhausted aged five years or less, after just three lactations, whereas a cow's natural lifespan could be 25 years. No surprise such stressed animals are susceptible to disease. The recent switch to maize as a key winter-feed is also suspected to be a factor, as maize is low in certain key minerals, such as selenium. Many farmers now supplement their animals' diet with selenium; some far-sighted ones have even done the same for the badgers on their land and appear to have kept TB at bay. Government scientists seem strangely reluctant to investigate this practical and positive farmer-devised approach.

Q Why is methane from cows such a problem? We all have to fart don't we?

A Social etiquette aside, we do all indeed have to fart! But some species are more flatulent than others – ruminant animals like cows and sheep especially. And we now have many of them to satiate the world's ever increasing appetite for meat and dairy.

In its report, 'Livestock's Long Shadow', the UN Food and Agriculture Organisation (FAO) describes animal production as "one of the top two or three most significant contributors to the most serious environmental problems, at every scale from local to global". Pretty alarming, given that the FAO also predicts that annual worldwide production and consumption of meat will more than double from the 229 million tonnes eaten at the beginning of this century to 465 million tonnes by 2050. The feed needed to fatten those animals has been calculated by agricultural writer Colin Tudge to be equivalent to all the food eaten each year by four billion humans – which was the world's total human population in 1970. Increased meat consumption has become a mark of affluence; consequently it is the poorest, hungriest people in developing countries who find their prime food-producing land taken up by crops grown to feed animals – much of it destined to produce meat for consumers in European and other wealthy countries.

Nearly a quarter of the world's land mass, excluding that still under ice, is grazed by livestock. But an even greater area is used to produce animal feed – in particular for the increasing numbers of pigs and poultry reared in intensive, indoor units. Most of the maize grown across the world and 95 per cent of the soya bean crop now goes to feed livestock. Soya production in Brazil, much of it on formerly forested land, has doubled to feed the expanding beef export market. Three-quarters of the cattle are stocked on what once was rainforest. The FAO has estimated that the clearing of forests and draining of wetlands for meat production releases 2.4 billion tonnes of carbon dioxide (CO_2) each year, which amounts to some seven per cent of global greenhouse gas emissions annually. With the rainforest goes not just its rich diversity of wildlife, but its capacity to absorb and lock away carbon dioxide from the atmosphere, and also its vital role (hence its name) in regulating water-cycles. The latter is compounded by cattle ranching's huge thirst for water – 11,000 litres of water are needed to produce the beef for just one quarter-pounder!

According to the FAO, farm livestock is responsible for one-third of all the world's methane (CH_4) emissions. Methane is a particularly potent greenhouse gas, around 20 times more damaging than carbon dioxide. Here's a fact that might make a determined petrol-head turn vegetarian –

the farts and belches of the world's cattle, sheep and goats create up to 18 per cent of all greenhouse gas emissions, more than the total global greenhouse gas emissions from all the world's vehicles. In the UK methane makes up 37 per cent of our overall greenhouse gas emissions, with agriculture and, specifically, farm animals and slurry from intensive animal units being the main source.

No surprise that vegetarians argue that theirs is the most planet-friendly diet. Environmental campaigner Jonathon Porritt agrees we should eat less, more-sustainably reared meat, describing excessive meat eating as one "of the gravest threats to the long-term sustainability of humankind" and "despairing at the selective blindness of politicians" to address it. Canadian ecologist Dr David Suzuki puts eating one meat-free meal a week high in his list of the, "ten most effective ways we can help conserve nature and improve our quality of life".

But if livestock are raised under principles of sound, sustainable husbandry in the right place – certainly for the most part outdoors – then moderate meat eating can make good ecological sense and contribute hugely to farming's productivity. Cattle and sheep should be grazing on hills, moorlands and wet grazing meadows, converting grass grown by the Sun's energy into milk, meat, hides and wool for human use – not competing for land better suited to growing arable crops and vegetables. In organic farming, livestock are an essential part of the fertility-building rotation of nitrogen-fixing clover and grass leys, adding the 'golden hoof' of their manure. Stocking rates for organic livestock are lower than in non-organic systems, and because the animals are less stressed, organic livestock (particularly dairy cows) live longer, productive lives – so reducing the amount of methane produced per steak or pint of milk.

If like Jonathon Porritt you're a 'conflicted carnivore', then the answer could be to eat less, better-quality, humanely reared, organic meat.

Q We're awash with water here. Why do we need to save it?

A Recent flood events across the UK could lead people to think that water shortages are not the problem. Indeed the Association of British Insurers has warned that the financial costs of flooding are likely to rise, with the annual flood bill across Europe increasing by up to £82 billion. But under worsening climate change, UK weather patterns are likely to flip-flop between cycles of flood and drought. Hotter, drier summers with rainfall declining by up to 50 per cent in the south and east of the country are predicted by the 2080s, along with warmer, wetter winters and sudden, heavy downpours occurring throughout the year.

In the UK, October is traditionally known to weather forecasters and reservoir managers as the start of the 'Water Year', but this seasonal marker is becoming less reliable. In 2003, due to the very late onset of significant rainfall, the routine replenishment of reservoirs and groundwater sources was down by 35 per cent. South-east England, with the greatest density of the UK's population, is also one of the driest regions – with ten times less water available per person than in Spain. Of course, there are much drier regions on our planet, which climate change will make even more arid. The United Nations Environment Programme estimates that desertification currently affects approximately 25–35 per cent of the world's land surface, an area covering 100 countries and home to about 1.2 billion people. Around 66 per cent of Africa's landmass is either desert or drylands, making it particularly prone to desertification. Many parts of the world already show signs of greater water scarcity – India, eastern Australia, Pakistan, China, Central Asia, Saudi Arabia, Egypt, North Africa, parts of southern Africa, southern USA and northern Mexico.

Despite being some of the driest countries on Earth, many of these countries are trapped on a crazy treadmill of using their best farmland to grow commodities like cotton, coffee and animal feed for export in order to earn enough foreign currency to buy foodstuffs from the world market to feed their people. Not only are they growing cash crops for export on land better put to growing food for their own people, but along with those commodities they are exporting their country's most precious, scarce resource – water.

The Aral Sea in the former USSR is the most notorious example of large-scale cotton-growing's thirst for water. The sea lost 60 per cent of its area and 80 per cent of its volume through four decades of unsustainable water extraction for irrigation to grow cotton. Cotton production and consumption is responsible for 2.6 per cent of global water use – it takes 1800 gallons of water to grow and process the cotton needed for just one pair of jeans!

Q Why can't we just adapt our farming methods and crops to suit the new climate created by global warming?

A The Thames Barrier, built in 1983 and predicted to be raised once every five to ten years, is now used six or seven times a year to protect London from flooding. Climate change scientists estimate that sea levels could rise between 26 and 86 centimetres above current levels in south-east England by the turn of the century. The exceptional and tragic coincidence of high tides and storms that led to the North Sea breaching Norfolk's sea defences in 1953, causing 307 deaths, could become a routine annual occurrence.

Declines in agricultural productivity and food security are predicted to affect many parts of Asia, as higher temperatures cause the retreat and melting of the Himalayan glaciers that store and release water into the Indus and Ganges – rivers that are the lifeblood of farming and food production through much of the Indian sub-continent.

Developed countries may have greater resources to contend with climate change, but they will not escape its impacts, either indirectly from the loss of imports of cheap foodstuffs from poorer countries or directly upon their own farming and food systems. With 57 per cent of the UK's best farmland lying below current sea levels and reliant on extensive drainage, pumping and sea defences, there are significant risks of widespread inundation. It is quite possible that arable farming to produce cereals could become unviable on 86 per cent of the East Anglian Fens and their famously super-productive black peat soils.

Q Why does it matter what the farmers do to the soil so long as we get food from it?

A Agricultural soils in good condition have the potential to act as natural sponges, soaking up excess rainwater and allowing it to trickle down over time to the key stores of our drinking water, the underground aquifers. The key to the functioning of this natural 'service' is a good, porous soil structure with plenty of organic matter, such as plant roots and leftovers from harvesting, like straw. But the organic matter content of UK soils has fallen dramatically as dependence on chemical fertilisers has replaced the traditional practice of using manure, compost and crop rotations, as well as ploughing in grass, clover and other crop residues to build fertility and improve soil structure. Loss of this organic matter means less carbon dioxide is locked away in our soils. Heavier farm machinery also compacts the soil, squashing soil particles together so it becomes less porous and water runs off the land, rather than filtering down through it.

Organic farming's reliance on crop and livestock rotations, composting and ploughing-in crop residues ensures far more organic matter in the soil, so maintaining soil structure and increasing its capacity for water retention.

During recent droughts in the US, organic farmers found their crops survived six weeks longer than their non-organic neighbours' shallow-rooted, fertiliser-fed plants – a critical period that enabled them to get their crops to harvest. US research carried out over 23 years shows organic soils hold 15–28 per cent more carbon than non-organic.

Soil scientists have calculated that a single teaspoonful of healthy soil can contain more bacteria, fungi, protozoa, nematodes and other species than there are people on the planet. In contrast, artificial fertilisers and pesticides significantly reduce the numbers of these microscopic soil inhabitants. Without chemical fertilisers poured on from above, organic crop plants send out deeper, more extensive root systems to find the nutrients they need throughout the soil profile. This matrix of roots, decaying plant waste and myriad creatures creates a well-structured, moisture-retaining, less flood-susceptible soil.

Artificial fertilisers and pesticides significantly reduce the numbers of these microscopic soil inhabitants. A 21-year field-trial in Switzerland comparing organic and non-organic farming showed dramatic differences in soil microbiology, with populations 85 per cent higher in the organically managed field than in that treated with artificial chemicals. The benefits for consumers of the food raised from this 'living soil' are confirmed by a growing body of studies. A review of available evidence in 2001 found that, on average, organic food contained higher levels of vitamin C and essential minerals than conventional produce. Research published in 2007 by the University of Newcastle showed that organic cows produce milk 50 per cent higher in vitamin E, 75 per cent higher in beta carotene (vitamin A) and two to three times higher in antioxidants, as well as with higher levels of omega 3 essential fatty acids.

The Soil Association, the UK's leading body promoting organic farming and food, was deliberately named to highlight the fundamental role of the soil in sustaining human health, as the organisation's founder Lady Eve Balfour said in her book 'The Living Soil' published in 1943, "My subject is food, which concerns everyone; it is health, which concerns everyone; it is the soil, which concerns everyone – even if they do not realise it". Her concerns about the negative impact on UK soils and their long-term fertility through the switch to chemical-based agriculture have been confirmed by government studies showing declines in minerals in UK fruit and vegetables of up to 76 per cent between 1940 and 1991.

Q Pesticides protect cotton crops from infestation and disease and surely increase yield, so we can have plenty of cheap clothing. Where's the harm in that?

A Around a quarter of all insecticides and ten per cent of the total pesticides used around the world go to grow non-organic cotton. For developing countries, it's more like half of the insecticides used. It takes roughly 150 grams (g) of agrochemicals to grow enough cotton to make just one t-shirt; 350g for a pair of jeans. After the crop is harvested, more than 8000 chemicals – many classified by the World Health Organisation (WHO) as 'hazardous' – may be used in processing the raw cotton into clothes and other items.

Some of those nasties inevitably remain in the finished material, ready to be absorbed through the body's largest organ, the skin. But it's those growing, rather than wearing, the cotton who are most exposed to the toxic chemicals and their effects. Headaches, nausea, eye problems and skin irritation are widely reported by cotton farmers and their families – including children – who work in the fields. WHO estimates that three million people suffer poisoning incidents and 20,000 people die in developing countries every year from pesticides, many of these linked to cotton production. In Benin in West Africa in 2000, 24 people died as a result of poisoning from cotton pesticides – including eleven children under the age of ten.

The history of pesticide use in cotton growing refutes the industry's claims that safety concerns have been exaggerated. Allegedly, just over 30 years ago the Swiss chemical giant Ciba-Geigy deliberately over-sprayed unprotected children working in Egyptian cotton-fields with the insecticide Galecron, to see how quickly they excreted the toxins. The company already knew that the active ingredient, chlordimeform, was a potential human carcinogen. The children later suffered diarrhoea, dizziness, head and stomach-aches, and other symptoms of chlordimeform poisoning. During the same period, company publicity in Europe warned parents to keep their children away from Galecron.

Cotton's dependence on large inputs of chemicals causes damage to soils and long-term losses to fertility. In a vicious cycle, poor farmers buy more chemicals in a desperate attempt to squeeze a crop from the ravaged soil, sinking further into debt. Organically grown cotton is different. The 25,000 farmers now producing it in 22 countries avoid this chemical-debt treadmill because, as well as growing cotton, their land can also be used to grow uncontaminated food for their families. Indian organic farmers plant pigeon peas between the cotton and make natural sprays from garlic, chili and the neem tree, all of which can be eaten or used as medicines, as well as protecting the cotton crop.

Chemical poisoning isn't the only price for cheap goods. Human-rights organisations, such as the US-based National Labour Committee, have successfully shone the spotlight on the dark side of the fashion industry: sweatshops where poor people, especially women, work in cramped conditions for minimal wages in vast factories in the 'free-trade zones' of Asia, Latin America and Africa, churning out jeans and t-shirts for designer brands and high street stores. But this compromising of human health and workers' rights runs all the way down the production chain to the farmers and labourers producing the raw material – cotton. Such callous use of human guinea pigs to test toxicity may be in the past, but the exploitation of children as labourers to produce cheap cotton for western markets is very much current.

In 2008, the major UK supermarkets were shamed into stopping sourcing cotton from Central Asian countries, such as Uzbekistan, where at harvest time the cotton fields are still filled with children (although this is against the laws). This disregard for children's welfare is not just due to corruption or failure to enforce international law, but driven by the crash in world market prices for cotton caused in part by the huge subsidies US cotton growers receive. In the past these have amounted to around $570 per hectare, with the top ten US cotton producers divvying out $17 million in direct payments from US taxpayers – cotton's nickname 'white gold' is accurate for those privileged few. Although under legal challenge by other cotton-growing countries such as Brazil, these subsidies are still available.

Q Food miles are an unavoidable evil. How can we go without bananas and oranges? And we want to support Fairtrade.

A The term 'food miles' was coined by Tim Lang, Professor of Food Policy, long-term food activist and thorn in the side of the agri-food business. It has been immensely positive in helping consumers understand the complex global food markets that fill supermarket shelves with an apparent endless cornucopia of exotic produce. When allied with its sister concept, Fairtrade, it gives consumers the information necessary to exercise their buying power environmentally and ethically. Global trade in foodstuffs, if balanced by environmental and ethical considerations, can bring great benefits to farmers and people in developing countries – while enabling western consumers' guilt-free purchases of globally-sourced exotics. I, for one, do not want to forego tea, coffee, chocolate, bananas or oranges. It might be technically possible to grow all those crops here in multi-storey heated glass-houses, but it would be ecological madness. The carbon footprint of tomatoes grown in southern Europe and trucked and shipped across to the UK is less than if those tomatoes were grown in heated greenhouses here (although using gas harvested from landfill-sites or electricity produced by wind generators to heat them could tip the balance).

That's not to say that countries shouldn't seek to grow as much of their own staple foodstuffs as possible. The wartime U-boat blockade of the convoys bringing in our food from our former colonies exposed the folly of excessive reliance on overseas imports. At the time, British farming had been left neglected and our farmers produced only about one-third of the nation's food. Although that's now risen closer to 70 per cent, our self-sufficiency in certain staple foods is again declining. This doesn't seem to worry politicians who believe the world market will continue to provide – apparently oblivious to the predicted negative impacts of climate change on some of the world's current largest food growing regions, including our own prime agricultural land in low-lying East Anglia.

Public awareness of food miles has boosted the burgeoning local food market, with over 500 farmers' markets and an even greater numbers of vegetable box schemes providing customers with more local, seasonal, unprocessed foodstuffs. But some locally caught, regionally sourced foods still make bizarre globe-trotting journeys, such as Scottish scampi being shipped to China for hand-shelling, then returning to Scotland for 'breading' before going on sale. The economic rationale is that cheaper labour costs in China outweigh the transportation costs – but they don't outweigh the full costs to the environment of this mad post-mortem prawn migration.

A critical step on the path to achieving sustainable food production, the concept of food miles now needs developing. Not just "How far has my food travelled?" but also, "How was it produced – using people and planet threatening chemicals or by more sustainable organic methods?". And "Who produced it and did they get a fair price?". Food miles plus Fairtrade plus organic offers consumers the holy trinity of sustainable, ethical food production.

Q Is there any evidence that chemicals used in growing food can harm us?

A Unfortunately, there's plenty of evidence that pesticides (literal meaning to 'kill pests') have unintended harmful effects on our health and the environment. It's no coincidence that 37 pesticides have been banned in the UK and 50 per cent of the pesticides registered across the EU have been phased out over the last 10–15 years. "Good – that means only safe, harmless-to-human pesticides remain in use", you might hope. But as recently as 2007, the pesticide Aldicarb was suddenly withdrawn from sale on human health grounds 15 years after the Soil Association and other environmental groups first called for its ban – and despite assurances from both the National Farmers' Union and agrochemical industry, just months before its withdrawal, that Aldicarb and modern pesticides generally were 'safe' and 'essential'.

Official surveys confirm that 40 per cent of fruit and vegetable items on sale in the EU are contaminated with pesticides – with one food item in 30 containing levels of pesticides above European legal limits. Not surprisingly, there's strong, ongoing public unease about pesticides. A recent EU-funded study of 25,000 European citizens, including 1334 from the UK, found that pesticides in food remain the number one food safety concern for people. Their concerns are neither groundless nor the result of green groups' 'scaremongering', but are supported by gathering evidence of the negative impacts of these toxic chemicals. Over the last 60 years, average male fertility has fallen by 50 per cent globally. While there are many possible factors to consider, that period does coincide with the increased use of pesticides in farming in the UK and worldwide. Incidence of breast, prostate and testicular cancers and male reproductive disorders, including undescended testes, has also increased in recent decades. Occupational exposures to some pesticides have caused reduced sperm counts and infertility in men, as acknowledged by the European Commission: "Long-term exposure to pesticides can lead to serious disturbances to the immune system, sexual disorders, cancers, sterility, birth defects, damage to the nervous system and genetic damage."

There are, of course, other views on the possible causes. An advisor to the Government's Advisory Committee on Pesticides stated at a recent meeting that there were a number of risk factors affecting male fertility "Such as too much time riding bikes, sitting down too much and wearing tight underpants".

Despite public concerns, non-organic farmers' reliance on pesticides is increasing. Over the past decade, use of insecticides in the EU and UK has more than doubled and all of the top ten insecticides used are classified as 'hazardous'. In the UK, according to Government statistics, 31,000 tonnes of pesticides are sprayed onto farmland each year to produce our food – half a kilogram for each and every one of us. Fortunately, a growing body of farmers share the public's concerns over pesticides and have switched to organic farming methods. For some this is a result of the direct effects on their own or their family's health. "In 1995 my wife became pregnant and was accidentally caught in a chemical pesticide spray on a conventional fruit farm. As a result of this (and a combination of contracting salmonella six months later) my son was born with a multitude of issues including long gap oesophageal atresia. He spent his first six months in intensive care at Great Ormond Street, and his first two years in and out of various hospitals around the UK. He has pretty much recovered now however he will always have a couple of issues that he will have to carry as a result of this episode. This was a bit of a wake-up call for me as to what we do to our food." Adam Wakeley, organic apple producer.

Q Farmers care for the land. How can farming be damaging the environment?

A Indeed, farming should surely be one of the 'greenest' of human activities, combining the natural resources of sunshine, rainfall and soil to raise crops and domesticated animals to produce our food. It can be close to that ideal, but unfortunately mainstream, modern intensive agriculture has shifted its reliance heavily onto products derived from the laboratory and dependent on large inputs of fossil fuels. In particular, artificial fertilisers, the 'magic mix' of nitrogen, phosphate and potassium, have boosted crop yields beyond farmers' wildest dreams, but require large amounts of oil and gas in their manufacture.

The 19th century chemist Liebig identified the basic minerals plants need by incinerating them and analysing the remaining ash. Liebig believed that by simply adding these back to the soil after cropping, fertility could be maintained indefinitely – thus setting in train a narrow, reductionist approach which ignores the key role the myriad of organisms in a healthy 'living soil' play in plant health and growth. Chemical-based farming treats the soil as an inert substrate, but good soil is made up of more than eroded geology. A spoonful can contain more bacteria, fungi, protozoa, nematodes and other fauna and flora than there are people on the planet.

Already implicated in polluting rivers, lakes and groundwater, nitrogen fertilisers also add to climate change.

The main greenhouse gas released through farming is nitrous oxide (N_2O), which is 200–300 times more potent than carbon dioxide. Agriculture is the single largest source of nitrous oxide in the world, thanks to non-organic farming's reliance on artificial nitrogen fertiliser. The overall energy cycle of making nitrogen fertiliser, transporting it from factory to farm and spreading it on the land, makes up between 37 per cent and 50 per cent of the total energy used by UK farming. Nitrogen fertilisers are also the largest source of carbon dioxide, the main greenhouse gas, produced by agriculture – with nearly 7 kilograms (kg) of carbon dioxide given off for every kilogram of nitrogen fertiliser made.

According to government figures, agriculture makes up about nine per cent of the UK's total greenhouse gas emissions and taking into account the whole food chain that rises to about 20 per cent. The food and farming sector accounts for over 31 per cent of the global warming potential of all products consumed within the EU, more than any other sector of the economy. So what we choose to eat can have more impact on climate change, for better or worse, than any other aspect of daily life.

Q Why do we need Fairtrade, when we already have free trade and the WTO?

A Like 'fair', 'free' is a good sounding adjective. Everyone wants to be free. What's the opposite? 'Imprisoned', 'enslaved'? So surely 'free' and 'fair' trade are brothers in arms, working to alleviate poverty and 'lift the boats' of developing countries, their economies and people's standard of living to that of our own in the developed world?

There's a world of difference between the two terms. The Fairtrade label, set up in the late 1980s by a coalition of over 20 international aid charities including Oxfam, recognised the opportunity for, and willingness of, concerned consumers in the developed world to help poor farmers in the South by paying a fairer price than set by global commodity markets. A price that not only met the farmers' costs of production but also enabled them and their communities to invest in vitally needed schools, sanitation and other infrastructure. The Fairtrade mark is now one of the best recognised ethical labels in the UK and has extended to whole towns, like Garstang in Lancashire, declaring themselves Fairtrade by promoting Fairtrade goods in cafés and shops, and local authority catering.

'Free trade' couldn't be further from this non-profit charity and grassroots-led initiative. Its champion is the World Trade Organisation (WTO), an unelected body set up after World War II to get the global economy turning again and focussed on cutting import tariffs and any 'protectionist' barriers to trade. Over the years WTO decisions have tended to favour the wealthy nations who set it up or, some would say, the international corporates who really pull the strings behind its façade of democratic control. Its decisions have always put trade first, whatever the consequences to the environment or human health. Controversial WTO rulings include that against the EU for restricting imports of tuna caught by methods that caused large-scale dolphin deaths. The US has also used the WTO to try and overturn the EU moratorium on US genetically modified (GM) crops and the longstanding ban on imports of US beef produced using growth hormones, prohibited across Europe on human health grounds.

It's a clue that the WTO and its declared agenda of 'free trade' and 'globalisation' is not universally popular that its meetings have attracted some of the largest public protests ever seen, bringing together labour rights activists, environmental groups, development agencies and small farming bodies, as well as more radical anarchist and anti-globalisation groups. The WTO meeting in Seattle in 1999, when armoured police tear-gassed and engaged in running battles with thousands of protestors, thrust the organisation and its formerly arcane business of trade deals into the media spotlight. The suicide of a South Korean farmer at the 2003 meeting at Cancún in Mexico showed the desperation felt by millions of small farmers in the South faced by international trade rules, world market prices, and indirect farming subsidies that favour farmers in the US and Europe and make it impossible for them to compete on the claimed 'free markets'.

No such protests or grim events surround meetings of the Fairtrade Foundation or its annual Fairtrade Fortnight. Fairtrade throws a vital lifeline to thousands of poor farmers across Asia, Latin America and Africa. Its very existence and name make it plain that most trade is far from fair. Strong consumer support gives sufficient market clout for big retailers to stock Fairtrade products as 'must-have brands'. The 'free traders' in Geneva and elsewhere may grind their teeth, but they won't change the minds of the people of Garstang.

Q Can we make poverty history through free trade?

A Encouraging poor farmers in the South to grow more export crops puts them in increased competition with other developing world producers doing the same, resulting in world market prices crashing further. This is good news for the global agrifood business processors and retailers, bad news for individual producers. It happened to Ethiopia, following World Bank advice to concentrate on growing coffee for export. Home to the original land-races upon which all commercial coffee varieties are based, Ethiopia would appear to have a strong 'comparative advantage' in that crop. Until recently, coffee made up more than 87 per cent of the country's total agricultural exports, around $6 of every $10 earned. But in 2003 world market prices for coffee plummeted, halving Ethiopia's export earnings from around $400 million to under $200 million. The global market had been destabilised through the emergence of Vietnam (a country with no previous history as a coffee exporter) as the second largest producer after Brazil, thanks to an investment programme encouraged by the World Bank, an institution, like the WTO, dedicated to 'free trade'.

Unfortunately, the WTO and its fellow-travellers seem to have succeeded in persuading some of those campaigning for a better deal for developing countries that there is now a real commitment to making international trade fairer and that more liberalised trade really is the best possible development path for the predominantly agricultural economies of many Southern countries. After all, isn't that what happened in the UK – agrarian revolution followed by industrial revolution?

A key objective of the 'Make Poverty History' campaign of 2005 was to open up developed countries' markets to imports from the South, particularly of agricultural goods, which those countries would seem to have a natural 'comparative advantage' to growing competitively. Somewhere along the process, the notion of an increase in 'fairly traded, fairly priced' food and textiles from the South got translated into simply 'more free trade'.

Opening up EU and UK markets to developing countries' produce sounds admirable, but doesn't bear much analysis. The EU is already the largest importer of African agricultural exports in the world, importing as much as the US, Japan, Canada, Australia and New Zealand combined – absorbing around 85 per cent of all of Africa's agricultural exports. The UK itself imports food from across the world, from the Caribbean, throughout Europe, across Africa, Asia and Australasia. What are we meant to do? Encourage people to consume more sugar, coffee, chocolate 'to help Africa'? Clearly absurd, and hardly in tune with Government targets to curb obesity and improve the nation's diet. The good intentions of those setting up the process to resurrect trade after World War II, bringing nations together through constructive commerce and exchange of culture as well as goods, was subverted by those merely interested in the profits, not the principles of trade. Just as the word 'free' has been subverted in the context of most poor countries to mean indebted, bonded and enslaved to international money markets, institutions and corporates.

This global swapping of agricultural products has been the mantra of the pro-free traders, who believe food or any other commodity should be grown or produced by the most efficient producers, where the conditions are most favourable. US Agriculture Secretary John Block, speaking at the global trade talks in 1986, dismissed those arguing for national food security: "the idea that developing countries should feed themselves is an anachronism. They could better ensure their food security by relying on US agricultural products, which are available, in most cases, at much lower cost." Specialising in growing animal feed or other cash-crops for export had some logic when other countries were producing sufficient human food crops for export. But in recent years, major exporting countries like the US, which grew 70 per cent of all grain on the global markets, have been diverting their crops to other uses or for their own food security. In 2007, one-sixth of America's entire grain harvest went not to feed people, but to make fuel for gas-guzzling US cars.

Q Can organic farming feed the world?

A Widespread use of chemical pesticides in agriculture has only been 'conventional' over the past 60 years. Yet the human population increased from ten million when settled farming first started around 10,000 years ago to over two billion by the 1930s. That 200–fold population increase was enabled and fed by traditional, effective, organic agriculture before the widespread use of agrochemicals. In the ensuing 70 years, the world population has more than tripled, but the increases in productivity brought about by industrial agriculture are far less dramatic than claimed – and indeed are further reduced when its reliance on vast inputs of oil, necessary to manufacture the fertilisers, fuel the machinery and from which many pesticides are derived, are taken into account.

A recent article in *The Economist* argued that producing the world's food organically would need so much land that we could "wave goodbye to the rainforest". The magazine shares the view of proponents of the agro-biotechnology industry that the only way to feed the world's burgeoning population is through whole-scale adoption of GM crops, animal growth hormones and continued use of chemical pesticides and artificial fertilisers. A genuinely altruistic mission to feed the world's poor and hungry is being hijacked as a marketing strategy for the GM industry. Take this plug from President Bush in 2003, "We should encourage the spread of safe, effective biotechnology to win the fight against global hunger". But global hunger is not caused by lack of land to grow food, rather because too much prime land is being used to grow food not for people, but inefficiently for fattening animals and increasingly for fuelling cars. The world's livestock eat one-third of all the cereals grown, with much of that animal feed produced in the South and exported to produce meat for consumers in the affluent North. Currently, the EU imports 70 per cent of all protein fed to its livestock. As the European Commission commented, "Europe's agriculture is capable of feeding Europe's people, but not of feeding Europe's animals".

During the Ethiopian famine of 1984, brought to the public's attention by BBC reporter Michael Buerk's harrowing pictures of starving families trekking to find food, the UK was importing large amounts of animal feed from Ethiopia – not much good for feeding humans, but the land it was grown on could have been. Cash crops don't just starve people, they sabotage their long-term food security by damaging the soils. The programmes funded by the World Bank and encouraged by the WTO for developing countries to specialise in cash crops, go hand-in-glove with promoting industrial, chemical-intensive farming methods. These leave a dangerous legacy. Ethiopia has the unenviable record of holding the largest stock of obsolete pesticides in Africa, some 3000 tonnes. Ethiopian farmers have a saying that illustrates the impacts of industrial farming: "The land is corrupted; it has acquired the taste for bribes. We have to bribe it with chemical fertilisers in order to produce anything." The FAO calculates that 80 per cent of pesticide poisoning cases occur in developing countries and that, "conservatively, over 60 million people (are) affected every growing season... an outrageously large proportion of those workers are children who are often more vulnerable to hazard and are poisoned at work".

The emphasis on cash-crop monocultures has also caused a catastrophic reduction in the diversity of food crops grown. Globally, the FAO calculates that 75 per cent of the world's agricultural diversity has been lost in the last half century. Some developing countries have the sense to step off this treadmill, instead prioritising feeding their people from their own farming and relying more on renewable human labour than environmentally and economically costly, finite fossil-fuel-based chemicals. Ethiopia, the former poster-boy of poverty and famine, is turning to its existing resource of 40 million predominantly small-scale, subsistence farmers to develop sustainable, organic systems, utilising traditional knowledge married with improved techniques such as composting, with astonishing results. Yields of the staple legume, the Faba bean have increased five-fold from 500 to 2500kg per hectare. Dr Tewolde Berhan Egziabher, Head of the Ethiopian Environment Agency, is the prime mover in shifting his country onto a sustainable farming path. "In a harsh climate and a largely agricultural economy we need to rediscover an approach to agriculture which supports long-term food security and protects soil fertility. Organic farming is the way forward for Ethiopia, and it is also an approach that can help to reduce the greenhouse gas emissions caused by mechanised farming and the petrochemical inputs in richer countries."

As a practical agriculturist in the South, Dr Tewolde commands our attention and should do more so than those seeking to flog fertilisers and find a new market for GM crops already rejected by European consumers. His views and observations are underpinned by Danish research presented to a UN Conference in 2007 that found that, in sub-Saharan Africa, a conversion of up to 50 per cent of agriculture to organic methods would be likely to increase food availability and decrease food import dependency. Organic yields do fall off to begin with, typically by only 10–15 per cent, but organic farming brings a bigger benefit in that poor farmers no longer have to buy or rely on expensive, imported fertilisers and pesticides. Alexander Mueller, assistant director-general of the FAO praised the research and said that considering that the impact of climate change will target the world's poor and most vulnerable, "a shift to organic agriculture could be beneficial".

It's hard to see many benefits for countries following the pro-GM, industrial farming cash-crop programme to the letter. Argentina is the third-largest grower of GM crops in the world. By 2000, more than 80 per cent of its total soybean acreage was GM. Yet by focussing on highly competitive commodity markets, its farmers' profits have halved, forcing over 50,000 off the land. As to 'feeding the world's poor', this model of agriculture hasn't prevented over 40 per cent of Argentinian children from being classed as 'undernourished' by the charity Caritas Argentina.*

Q Is growing crops for biofuel the answer to peak oil?

A Superficially, biofuels sound great, literally 'growing green energy' annually to displace fossil fuels laid down over geological timescales. On closer scrutiny, the arguments in their favour unravel. Just look at who's pushing them: car manufacturers, GM companies, and former barley-barons at the National Farmers' Union HQ hoping for a bumper harvest subsidised by the taxpayer! And being an industrial, rather than a food crop, one they can grow without consumers worrying about pesticide residues. No surprise then that GM companies are seeking to use the guise of 'green' energy crops as a Trojan horse to get GM plants into UK and European fields. Because the main crops proposed – oil-seed rape for biodiesel and wheat or maize for bio-ethanol – still rely on large inputs of artificial nitrogen fertiliser, the true energy ratios are dismal. An energy cycle audit of biofuels carried out for the government by Sheffield Hallam University demolishes the biofuel lobby's claims: for every £1 invested, biodiesel saves at best 5kg of carbon dioxide, compared to 35kg per £1 for fitting homes with more efficient, condensing gas boilers, or 500kg per £1 for lagging household lofts properly.

Internationally, the race to feed cars rather than people from crops such as palm-oil and sugar-cane is exacerbating hunger and destroying the very ecosystems that are key to locking away vast amounts of carbon dioxide – the world's tropical rainforests. To compensate for the carbon released from forest cleared for biofuel production would take hundreds of years of palm-oil substitution. And palm-oil plantations can never provide the habitat for endangered animals like the orang-utan, or homelands for the displaced tribal peoples.

The US used to supply 70 per cent of all grain on the global market. 'Used to', because since 2006 America has been diverting one-sixth of its grain harvest into making biofuels in a desperate effort to reduce dependency on Middle Eastern oil. Bio-ethanol plants for turning wheat and maize into fuel are being constructed across the Corn Belt at a staggering rate – over 55 have been built or planned in Iowa alone. Consequently, wheat prices in the US have risen 50 per cent over the past two years and grain exports have also been slashed. The US Department of Agriculture estimated that, of the extra 20 million tonnes of wheat grown globally in 2006, 14 million tonnes would go to 'feeding' American cars with just six million tonnes left for the world's growing numbers of hungry people. The grain needed to make enough bio-ethanol to fill a typical gas-guzzling SUV's 110-litre tank would feed one person in the South for a year. No wonder Jean Ziegler, the United Nations' independent expert on the right to food, called the race to grow biofuels "a catastrophe for the poor" and a "crime against humanity".

Not all of the proposed biofuels are bad. Biomass crops grown for burning to create space, heating or generate power, such as short-rotation willow and elephant grass, have the best potential (along with biogas from animal wastes). Willow can produce 30 times the energy used in growing it, compared to at best the 2–4 times energy-out for energy-in claimed for biodiesel from oil-seed rape. Through photosynthesis, biomass is the product of Sun, soil and water. But the potential to capture more 'free energy' from the Sun, while avoiding locking land away from more sustainable, rotational organic farming for food production, is enormous. Friends of the Earth have calculated that if we could capture just one ten-thousandth of the Sun's energy that hits the Earth, we could supply more energy in a year than we get currently from burning all fossil fuels. Rather than millions of hectares of productive farmland under industrial-scale fuel crops, wouldn't it be better if millions of hectares of urban rooftops were covered in solar heating panels and electricity-generating photovoltaic panels?

Q Can big multinationals produce Fairtrade, ethical and organic food for all?

A There isn't yet an organic Mars bar, but the Mars Corporation does own the organic food brand Seeds of Change, which began life in 1989 as a radical, seed-saving organisation. The story goes that an intern working at the organisation research farm in New Mexico suggested his family firm could help create and promote food products made from Seeds of Change's seeds. The intern's name was Mars. According to Seeds of Change founder Howard-Yana Shapiro, the pairing between the corporate food giant and seed activist organisation is entirely virtuous – marketing muscle allied to authenticity and field research, which includes Fairtrade deals with the communities in the South providing the source seeds. A similar story goes for Green & Black's, the pioneering chocolate brand set up by Soil Association chairman Craig Sams and his wife Jo Fairley. The company's flagship bar, Maya Gold, was the first product to bear both the Soil Association's organic symbol and the Fairtrade logo. In 2005, Green & Black's was bought by Cadbury Schweppes, resulting in the brand being downgraded by *Ethical Consumer* magazine from third-best ethical chocolate bar to 12th, just above Kit-Kat made by Nestlé, a longstanding whipping-boy for ethical consumers for pushing breast-milk substitutes to African mothers. But for Craig Sams, Cadbury Schweppes has brought in the necessary investment to develop the brand, without compromising its principles; thousands more hectares of land are now under organic production and the small cacao growers still enjoy Fairtrade deals. As Sams says, "Some people have said that ultimately we will have engineered a sort of reverse take over, on a cultural level, of the world's largest confectionery company".

But not all big business involvement in organics is in keeping with the movement's grassroots origins and principles. Influenced by the lobbying power of large food corporations wanting an easy-in on the burgeoning organic market in 1997, the US Department of Agriculture, which oversees the standards set for organic food and farming in the USA, sought to allow irradiated food and even GM organisms under the then-new federal organic food regulations. Over 200,000 US citizens, activists and organic farmers and consumers wrote in to protest and staved off such a cynical lowering of the bar. An internet search on corporate organics pulls down over 700,000 entries, the vast majority from the US, indicating the power struggle raging for the soul and principles of the organic movement. The recent diktat by the EU that a generic Euro-organic logo must be printed on all European organic produce is another worrying example of organic principles being bent to suit corporate interest.

The stated objectives of harmonising markets and maximising global competitiveness are crucial for global agribiz, but run completely counter to the values of the pioneers of the organic movement, as well as to the recent resurgence in locally produced food. As the Soil Association founder Lady Eve Balfour wrote 60 years ago, "If fresh food is necessary to health in man and beast, then that food must be provided not only from our own soil but as near as possible to the sources of consumption. If this involves fewer imports and consequent repercussions on exports then it is industry that must be readjusted to the needs of food. If such readjustment involves the decentralisation of industry and the re-opening of local mills and slaughterhouses, then the health of the nation is more important than any large combine."

She would have been dismayed at the idea of some standardised 'Euro-organic sausage', as would most present-day organic consumers who want to buy regionally distinctive sausages from Norfolk, Cumberland or Schleswig-Holstein, made from locally reared, traditional breeds of pig.

Q Is modern oil-based farming more energy efficient than traditional farming?

A During the last hike in oil prices caused by the Arab–Israeli war of 1973, economist Gerard Leach crunched the numbers of oil-dependent modern agriculture compared to true horse-powered pre-World War II farming. Initially, machine-powered food production out-yielded horse-power by 200 food energy equivalents out to each put in, which would seem obvious, given that about a quarter of all farmland had to be set aside for growing hay and other fodder for the horses.

But once the full costs of the oil used to manufacture, power and maintain the tractors and produce the fertilisers was put into the equation, horse-powered farming became twice as energy efficient than diesel-dependent farming. No one's suggesting wholesale return of shire horses yet, although the double challenge of curbing climate change by reducing our fossil fuel use and the rising price of oil as remaining reserves become scarcer and harder to extract, mean that food production like everything else can no longer rely on 'cheap oil' to fuel it.

Seventy-two per cent of agriculture's use of energy is indirect use of energy due to the reliance on inputs of fertilisers and other agrochemicals. Organic farming is more energy efficient than non-organic farming, due to the non-use of fertilisers and non-use of other inputs. A recently completed life cycle analysis of ten organic and non-organic sectors for the government found that organic farming on average requires about 25 per cent less energy to produce the same amount of food as non-organic food production.

Q Even organic farmers use chemicals to grow their crops, don't they?

A Organic critics often leap onto the fact that organic farmers can use a small number of pesticides. "Ah ha!" they crow, "Not so pure then!" But the fundamental principle of organic farming methods is to avoid the use of artificial inputs and chemicals wherever possible. By relying instead on rotations of crops and livestock, organic farmers break the cycles of disease and pest attack that can bedevil non-organic growers. Artificial chemical fertilisers cause rapid, lush growth and weaken plants' cell structures, making them a magnet for bug and fungal attack – so necessitating further use of chemicals.

Sir Albert Howard, a pioneering agriculturist and inspiration behind the founding of the Soil Association, described pests and disease, as "Nature's professors of agriculture" and "an integral portion of any rational system of farming". He meant that any excessive, economically damaging pest or disease attack showed that the farming system was out of balance. While seeking to sustain 'balance' naturally, organic farmers can use a limited number of pesticides as a last resort.

General organic standards allow up to seven pesticides, but the Soil Association only permits four, mainly to control insect pests and fungal attack. The four allowed under Soil Association standards are either of natural origin, (rotenone from the root, derris, and soft soap) or derived from simple chemical products (copper and sulphur). In contrast, non-organic 'conventional' farmers can use over 300 synthetic chemical pesticides. Annually over 31,000 tonnes of these get sprayed onto UK farmland, compared to around ten tonnes of the four pesticides available to the Soil Association's farmers – which amounts to just 0.03 per cent of the total pesticides used in agriculture. If all UK farmland switched to organic there would be a 98 per cent reduction in pesticide use.

Because organic farmers rely on rotations of crops, and generally include livestock in producing manures, correspondingly a greater variety of tasks and skills are needed to work the system, and that means more input of human labour. Is that such a bad thing? Conventional agronomists would say so. The drive towards ever greater 'efficiency', with farms shifting from mixed-farming to specialising in either crops or livestock and relying on machinery and chemicals in place of people, has caused the loss of hundreds of thousands of farm jobs – averaged out over the past 60 years, 37 farm workers have left the land every day. A 2007 study found that on average 32 per cent more jobs were provided by organic compared to non-organic farms. Also, on organic farms the average age of those farming was seven years younger than for the industry as a whole – 49 rather than 56. It also found that there were more women active in organic farming and when questioned about the future for themselves and their children, organic farmers were generally more optimistic about the future and happier!

The replacement of people by machinery and chemicals has undoubtedly ended some tedious, back-breaking jobs, but it has also contributed to the loss of rural infrastructure, as all the services from local abattoirs to village shops, pubs and post offices that depended on those living and working in the countryside decline in tandem. Four out of ten parishes in rural England now have no shop or post office, six out of ten have no primary school and three-quarters lack a bus service or any form of health clinic. 'Efficiency' comes at a price.

Q Does it take more energy to maintain organic crops because they are more labour intensive?

A It depends how you define 'energy' and how far along the food production and distribution food-chain you look. Typically, organic farming has been calculated by government scientists to use just over a quarter less energy to produce the same amount of food as non-organic agriculture. This is mainly because organic farmers do not rely on artificial chemical fertilisers, which require huge quantities of oil and gas in their manufacture, as well as transport to and application on the farm. For every tonne of nitrogen fertiliser made, nearly seven tonnes of carbon dioxide is produced.

In contrast, organic farmers 'capture' nitrogen from the atmosphere using the Sun's energy to grow clover. Through photosynthesis, clover 'fixes' nitrogen naturally in the soil with the help of soil bacteria called rhizobia. The simple clover plant, much beloved of bumble-bees, provides excellent grazing for cows and sheep and is the beating heart powering most organic farms. Clover can fix 200kg of nitrogen per hectare, more than the average 190kg per hectare non-organic 'conventional' farmers use to grow their wheat, oil-seed rape and potatoes. With fertility further boosted by the grazing animal's manure, clover prepares the ground for a series of 'extractive' crops such as potatoes or wheat that take up and use the nitrogen.

While some energy analysts claim that organic livestock farming is less efficient in basic terms of feed energy-in to food energy-out, this fails to consider the full life cycle of the relative farming systems. Cramming 40,000 chickens into a windowless shed or running an intensive pig unit with similar numbers of animals confined indoors may be very efficient at producing meat in a short period of time, but that efficiency is at the expense of animal welfare, and brings greater risks to human health from routine use of antibiotics, as well as pollution from the vast quantities of slurry generated by the confined animals. Slurry, a sloppy mix of urine and manure, produces more of the greenhouse gas methane than the more solid manure of outdoor-reared animals.

But there's no doubt that organic farmers will need to further reduce their 'carbon footprint'. Some like Ian Tolhurst, who grows vegetables organically in the Thames

Valley, are already far down this path. Tolhurst Organics feeds 400 families through its organic veg box deliveries. Yet its total carbon footprint is less than the average four-person UK family, making it 90 per cent more energy efficient and climate-friendly than if that same food came from a non-organic farm, dependent on large inputs of diesel and chemicals, and bought from a typical supermarket with its multi-miled processing, packaging and distribution systems.

Consider the fossil-fuel energy requirements of the most 'efficient' US farms. To produce, process and distribute a year's worth of food for the average US citizen requires 1800 litres of diesel fuel. On average in the US, ten calories of fossil-fuel energy go to produce each calorie of food energy. For some crops, such as canned corn, the ratio jumps to an incredible one hundred to one. Promoting such food production models as the means to ending hunger and poverty is totally unrealistic and contradictory. Developing countries couldn't afford the oil and, even if they could, the world can't afford the climate impact from the added carbon dioxide produced.

When the Soviet Union collapsed, so did Cuba's imports of oil and agrochemicals, so the country had no option but to achieve greater self-sufficiency. From a higher application per hectare than the US, Cuba cut annual pesticide use from 20,000 tonnes nationally in 1989 to around 1000 tonnes today. Crop rotations and the use of green manures and development of bio-fertilisers from naturally occurring soil bacteria and fungi have helped. But the statistic that has the most significance in a post-peak oil world is Cuba's increased reliance on human labour: some 15–24 per cent of the population have become involved in agriculture since the early 1990s. Currently, in the UK less than two per cent of the population works in agriculture, compared to 40 per cent in 1900.

Q Poverty and illness in the developing world is unrelated to climate isn't it?

A It's one of the cruel ironies of climate change that its impacts are predicted to hit developing countries the hardest, despite these being the lowest emitters of greenhouse gases per head of population. And because of a lack of supporting infrastructure and readily available emergency services, people in low-income countries are four times more likely to die in natural disasters than people in wealthier countries.

Scientists predict that a warmer atmosphere will cause more intense, sudden downpours, with consequent floods and cyclones. A few 'contrarian' scientists still question the direct link to human-exacerbated climate change, but weather-induced disasters like Hurricane Mitch in Nicaragua, the super-cyclone and coastal inundations in Orissa, India and the ongoing floods in China, Bangladesh and Mozambique are all typical of the extreme weather events climate change is predicted to cause. According to Oxfam, disaster losses globally increased from $71 billion in the 1960s to $608 billion in the 1990s. Over 10,000 Indians perished during the super cyclone of 1999 that hit the eastern Indian state of Orissa, yet these victims of our increasingly violent global weather quickly faded from the consciousness of the western media and public conscience.

At the climate talks in Bali in 2007, a delegate from Papua New Guinea, challenged America to "either lead, follow or get out of the way", shaming the world's largest emitter of greenhouse gases into acknowledging the facts of, and human responsibility for, dangerous climate change. Per head of the population the US produces five times more carbon dioxide than China, 20 times more than India, and 200 times more than Mozambique.

Wasting Away

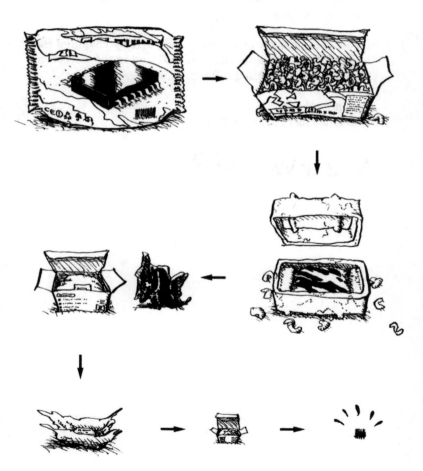

Waste Watch is a leading environmental organisation working to change the way people use the world's natural resources. Our vision is a less wasteful society, and we believe everyone has a part to play in reducing waste and living more sustainably.

We are a national, independent, not-for-profit organisation and a registered charity. We advise, educate and inform people on ways to reduce all forms of waste – from the rubbish we throw away and the energy we needlessly use, to the water we pour down the drain.

By making changes to our daily lives and by changing the world around us – at home, in school, and in the workplace – we can all make a big difference.

Steven Webb
Monitoring and Evaluation Officer

Change the world around you

Q Why recycle when it takes more energy to collect and sort than to make new?

A It is a common myth that the collection and sorting of recyclable materials uses more energy than manufacturing products from new materials, making recycling a pointless and energy-intensive exercise. The reality, in most cases, is that manufacturing using new or 'virgin' materials uses more energy than manufacturing from recycled materials. For example, manufacturing an aluminium can from recycled aluminium cans uses only five per cent of the energy used to manufacture a can from virgin material.

And the benefits gained from recycling aren't just in energy saving. Manufacturing products from recycled rather than raw materials avoids the environmental costs associated with the extraction of raw materials. These include the pollution of groundwater from mining metals, caused when rainwater percolates through soil contaminated with mining waste, or the loss of biodiversity when species-rich natural forests are replaced with plantations of a single species of tree to provide pulp for the manufacture of paper.

And what about disposing of materials? If we don't recycle our waste, we need an alternative waste disposal route – in the UK this has traditionally been landfill or incineration. There are many problems with landfilling waste. Liquids leach from the buried waste and find their way into watercourses, damaging ecosystems and contaminating drinking water. Waste in landfill degrades anaerobically (in the absence of oxygen), producing methane (CH_4), which is a greenhouse gas approximately 20 times more potent than carbon dioxide (CO_2). Human-made emissions of greenhouse gases are the main cause of climate change. Although recent legislation means that modern landfill sites are much more environmentally friendly than they used to be, the emissions from them cannot be stopped completely so the landfilling of waste still poses a significant threat to the environment.

In the absence of viable alternatives, the only other waste disposal option is incineration. Some materials, such as plastics, have a high calorific value and can be used to generate energy as heat in waste plants, but much of our waste has a low or even negative calorific value – fuel needs to be added to the waste to enable it to burn. Combined heat and power incineration plants are in their infancy in the UK but there is potential to develop these as efficient waste disposal systems. The heat and energy generated can be fed into the grid or used locally to power homes or industry.

There are also safety issues linked with incineration. People are concerned that toxic gases may escape, especially when temperatures are not exactly maintained. Also, very fine particulates may be discharged – these are extremely pernicious contaminants that can pass from the atmosphere into animals, including humans. The toxic gases emitted include dioxins, which are carcinogenic (cancer-causing) and oestrogenic (leading to lowered fertility in males), as well as large quantities of greenhouse gases, not normally classed as toxic, such as carbon dioxide. On top of this, the ash (and fly-ash from the filters) may contain poisonous compounds containing heavy metals, which have to be disposed of. These residues can account for about 25 per cent of the original waste. Many incinerators are blamed for reducing the air quality in the local area and causing headaches and general malaise, though the link between incinerators and illness has not been proven.

Recycling collection and sorting systems are now widespread, and this industry growth, coupled with technological improvements, is helping to make recycling more efficient. The environmental gains from recycling are more obvious for some materials than for others. By recycling everything you can, you are supporting your local recycling industries and by buying products with recycled content you can help to create a larger and more efficient UK recycling industry and avoid incineration.

Q We have more recycling than we can cope with, don't we?

A We don't have more recycling than we can cope with! This myth may have been started by people who have seen the piles of recyclables at Materials Recovery Facilities (MRFs) where materials are sorted and then baled for reprocessing.

MRFs sort and bale materials from a variety of sources. Any change between sources usually means a change in the set-up of the system, which takes up valuable sorting time. So MRFs stockpile recyclables, to keep any set-up changes to a minimum and therefore be as cost-effective as possible. Sorted materials are baled and stored until there is a sufficient quantity to send for reprocessing. Often a minimum quantity of a material must be collected and baled before it is purchased for manufacture into new products.

There have been some problems in recent years that may have caused an MRF to be out of operation for a period of time. In some cases this has led to a build up of materials that the MRF or the recycling collection companies have been unable to store until the plant is able to clear the backlog. In these rare cases, materials collected for recycling have been disposed to landfill or incineration. These problems have been very rare, but widely publicised. Recycling is a relatively new industry in the UK and, like any new industry, it has had inevitable teething problems. As the UK recycling industry develops, operational problems are less likely. If one MRF is out of operation, there is now more capacity to clear any backlog, and less need to resort to landfill.

Q Is there any point in driving to the recycling centre and burning fuel?

A It is worth checking with your local council what materials can be recycled through your doorstep recycling collection service – you may not need to transport your own. If you do have to travel to a recycling centre, you could always walk or cycle. By making frequent journeys to the recycling centre by bicycle or on foot you will help maximise the environmental gains from recycling your waste, at the same time improving the local air quality and keeping fit in the process.

There is no point in driving the car to the recycling centre and burning fuel if this is the sole purpose of your journey. The environmental benefits from recycling are greater for some materials than others, but if you are driving a vehicle even a short distance, the environmental impact of your vehicle's emissions will certainly reduce the environmental gains that your recycling produces. Instead combine your trip to the recycling centre with one that you were going to make anyway, for example, on your way to visit friends or family, or on your journey to school or work.

Many supermarkets now have recycling banks in their car parks. Combining your weekly shop with your recycling saves an extra trip to the recycling centre. Even if you don't shop at the supermarket, the recycling banks may be your closest recycling centre. If you have space, collect and store your recyclables for a longer period of time before you take them to the recycling centre. This will reduce the number of journeys you make to recycle and therefore reduce the environmental impact of transportation.

Q Doesn't recyclate all get shipped overseas, mostly to China, so generating lots of carbon emissions before it gets dumped anyway?

A It is not true that all of the materials that are collected for recycling in the UK get shipped overseas. The UK market for recyclate (recyclable materials that have been sorted and baled for reprocessing) is still developing. The UK is now collecting vast quantities of recyclate, but still lacks the infrastructure required to reprocess all the recyclate and manufacture recycled-content products on a large scale. Countries such as China have a huge demand for resources, and established markets for recyclates. They are able to efficiently turn our recyclate into new products. Export of recyclates is big business, and 237,753 tonnes of plastics were exported in 2005.

The carbon emissions associated with shipping recyclate collected in the UK to China are very low. Since the UK imports far more products than it exports, this means that many ships return to China with empty containers on board. The ships have to carry a minimum weight in order to float properly, and they would carry water if recyclate wasn't available. As there is so much space available on ships returning to China, transporting recyclate makes both economic and environmental sense.

It is highly unlikely that recyclate is shipped thousands of miles and then dumped. The recyclate is far too valuable a resource to simply be buried when it arrives, and strict legislation prevents the UK exporting materials that are not of a sufficient quality for reprocessing.

A larger UK market for recyclate would produce major environmental benefits. If we produced more products made from recycled materials in the UK we could reduce our imports and the carbon emissions associated with transporting new goods over long distances would be greatly reduced.

Remember that you are not truly recycling until you purchase products made from recycled products and 'close the loop'. By purchasing products with recycled content produced in the UK, you will not only help to save valuable resources – you will also help to develop and strengthen the UK recycled-products industry and reduce the quantity of recyclate being exported.

Q What's the fuss? Everyone recycles now anyway.

A It would be great if everyone recycled, but unfortunately they don't. Many people who do recycle don't use the available facilities to their full potential. Many households who have a doorstep multi-material recycling collection scheme only recycle two or three of the materials accepted, when they could recycle a much wider range. On average, participation in recycling schemes stands at approximately 65–70 per cent of households. This low participation rate, coupled with the fact that many households don't recycle the full range of materials they could, means that recycling schemes are not nearly as comprehensive as they could be.

Although municipal waste is a relatively small percentage of the total amount of waste produced overall, it is a highly significant proportion. It contains large quantities of glass, metals and plastics – materials that are in high demand by reprocessors. It also contains large quantities of garden and food waste, which cause pollution problems when buried in landfill (see page 126). The elimination of biodegradable waste entering landfill through home composting and doorstep collection will reduce greenhouse gas emissions.

Latest figures reveal that the UK recycles or composts just over 30 per cent of its municipal waste. Over 50 per cent of municipal waste has the potential to be recycled or composted, indicating that nearly half our recyclable waste ends up as landfill or is incinerated. On average, each UK household produces over one tonne of waste annually and sends almost three-quarters of that to landfill or incineration. Some other countries do better: Austria and The Netherlands recycle around 60 per cent of their municipal waste. The UK has a long way to go to catch up.

If you would like to start recycling but feel a bit overwhelmed by all the options, why not start by recycling only one or two materials such as paper or glass? You will soon get into the recycling habit and find that it is simple and doesn't take any extra effort. You can start introducing new materials as your recycling knowledge and confidence increases.

Q The energy it takes to wash and dry nappies makes disposable nappies an acceptable alternative, doesn't it?

A We all need nappies for the first couple of years of our lives and all nappies have an impact on the environment, whether they are disposable or reusable. With disposables, the main environmental impact comes with final disposal. Around eight million disposable nappies are thrown away in the UK every day – that's almost three billion a year. Almost all discarded disposable nappies end up in landfill and we don't know how long they take to degrade. The first disposable nappy ever to be used is probably still sitting in landfill somewhere!

Disposable nappies cause significant problems in landfill because they absorb liquids from the general waste and balloon to several times their original size and weight, taking up valuable landfill space. They also contain human waste, which combines with liquids in the general waste to create a toxic cocktail that can be very damaging to the environment if it escapes from the landfill site.

The main environmental impact from reusable, or 'real' nappies is from energy consumed when they are washed and dried. By following the four simple steps below, you can keep the environmental impact to a minimum.

- Use an energy efficient 'A' rated washing machine.
- Wash soiled nappies in a 60°C wash; there is no need to boil them.
- Wash wet nappies and liners in a cooler wash with the normal laundry.
- Don't use a tumble dryer; nappies should be dried on a washing line or clothes horse where possible. There is never any need to iron nappies or linings!

For most people, reusable nappies conjure up images of nappy rash, smelly buckets and white towelling squares that you need to be an origami expert to fold. This could not be further from the truth. Reusable nappies now come in bright colours with modern designs, in various sizes, and are extremely easy to use. The linings in the reusable nappies make it easy to dispose of waste down the toilet where it is properly treated, rather than into a landfill. And if you really can't face the idea of washing nappies, there are a growing number of reasonably priced nappy laundering companies in the UK who will take away your soiled nappies and deliver freshly laundered and ready to use ones. The economies of scale associated with nappy laundries make them extremely efficient in terms of energy and water use.

You can buy all the reusable nappies you need for your baby's nappy-wearing life from as little as £50, including waterproof covers. The same amount of money would only keep a baby in disposables for around two months. Many people are put off buying reusable nappies by the initial outlay, but there is some good news. Local councils are keen to divert disposables away from landfill, so many provide grants to enable families on low incomes to purchase reusable nappies. They can also provide expert advice and information on local reusable-nappy schemes.

Q Zero waste is a throw-away concept; how can it be possible?

A The term 'zero waste' originated from the Japanese industrial concept of 'Total Quality Management' (TQM). TQM ideas such as 'zero defects' have been enormously successful and have enabled major Japanese manufacturers to produce electronic goods to such a high standard that as few as one unit in a million is defective.

Turning the concept of TQM towards waste management deals with the whole life cycle of products, rather than simply trying to find 'end-of-pipe' solutions. We can stop viewing our waste as something useless to be disposed of and start viewing waste as a resource. Zero waste makes us think about why waste is produced in the first place and how we can prevent it.

The goal of zero waste is ambitious, but we can go a long way towards achieving it. The entire life cycles of all the products and services we use need to be carefully thought out, to ensure waste is kept to a minimum. Everybody involved in the life cycle of products, from materials extraction through to final disposal, is responsible for making zero waste work.

Responsible production is one of the key requirements for achieving zero waste. Producers need to make sure that as little as possible is wasted during the production stage. They need to look at the waste produced and find ways of minimising it, perhaps by using different materials or more efficient production processes, and by identifying ways of reusing or recycling any leftover materials. They also need to look at the materials they use to produce goods. Can they use less of a material, more environmentally friendly materials, recycled materials or materials leftover from other products? Can the product and packaging easily be recycled at the end of its useful life? Producers also need to look at other non-material waste, such as energy and water use, and discharges into the environment such as emissions from factory chimneys.

Consumers must play a part too. We need to think about the environmental impacts of our purchases. Do you really need to buy this product? Can the packaging easily be recycled? Can the product be recycled at the end of its useful life? This assumes that you only buy new products. By buying second-hand goods, consumers avoid the waste associated with the manufacture of new products and at the same time prevent unwanted goods from becoming waste.

Final disposal is the most visible part of a product's lifecycle, and this is where we will know if we have got all of the other parts right. If everything has gone to plan, we will have significantly reduced the volume of waste that goes for final disposal. For the waste that we can't reuse or recycle, we need effective treatment processes to ensure that we get some value out of it, such as energy recovery.

Q I don't compost my waste. It goes into landfill and rots down anyway. What's the difference?

A Many people believe that when waste degrades in landfill it does so in the same way as in a composter, but this isn't the case. Composting is the most efficient way of disposing of biodegradable waste and is beneficial on both a local and a global level.

When biodegradable waste rots in a landfill site, it is locked away from oxygen in the atmosphere. Anaerobic organisms that can tolerate a lack of oxygen break down the waste and this process produces methane. However, when waste rots in an oxygen-rich environment such as a composter, the organisms that break down the waste produce carbon dioxide, a greenhouse gas less harmful than methane. Biodegradable waste in landfill can also degrade into a toxic mix of chemicals that leach from the waste and enter watercourses, polluting drinking water and damaging aquatic ecosystems.

By composting our biodegradable waste, we not only help reduce greenhouse gas emissions from landfill we also return nutrients to the soil. Adding compost to the soil improves the structure and fertility, and increases its ability to absorb carbon dioxide.

There are also strong economic arguments against sending biodegradable waste to landfill, and the EU Landfill Directive aims to reduce the amount that ends up there. Under the Directive, UK Waste Disposal Authorities have a limit on the tonnage of biodegradable waste that they can send, if they exceed their allowance they are fined for every excess tonne – currently £150 per tonne. This may not sound much, but for authorities handling thousands of tonnes per year, failing to divert biodegradable municipal waste from landfill could prove extremely costly to the council tax payer. The fines for exceeding landfill targets all come out of the public purse, and this money could be put to far better use elsewhere.

It is always better to compost your organic waste at home by putting it in a composter or wormery – you don't need a lot of space for this. A simple compost bin will sit in a corner of a small garden and a wormery is normally small enough to fit on a balcony or terrace. In wormeries, worms break down the waste; they generally work faster than conventional composters and produce superior compost known as 'black gold', and a liquid known as 'worm tea'. Both are excellent soil improvers.

For biodegradable waste that can't be composted, such as meat or fish leftovers, many local authorities offer a kitchen waste collection service. The collected waste is composted in a commercial composting facility which reaches higher temperatures than household composters, killing off any harmful pathogens. Garden waste can usually be taken to your local Household Waste and Recycling Centre where it is collected and taken to an industrial composting facility. Community compost schemes are also springing up across the UK, funded in part by local authorities.

Q I'll start recycling when supermarkets stop using unnecessary packaging.

A Packaging can provide necessary protection to perishable goods, ensuring that they keep fresh and in good quality for when we want to use them. However, there are numerous examples of over-packaged goods in our supermarkets.

Packaging waste makes a significant contribution to household rubbish, and is a major source of litter. Some packaging companies have worked to reduce the environmental impact of their packaging by moving to lighter weight products. Glass bottles and steel or aluminium cans are now approximately 30 per cent lighter than they were in 1980.

Most local authority doorstep recycling schemes collect common packaging items: cardboard, glass, plastic bottles and cans; additional recycling facilities are often available at supermarkets and local recycling centres.

There is legislation against excessive packaging: the 1994 European Packaging Directive ensures that the packaging industry meets strict requirements and that packaging used is not excessive. To date there have been very few prosecutions under this legislation, but widespread concern remains among consumers that producers and retailers continue to sell over-packaged products.

The best way to solve the problem of packaging disposal is to avoid it in the first place. Try to buy locally produced goods, as these don't require as much packaging as products that have to travel thousands of miles; you will also be reducing the carbon emissions associated with the transportation of your goods. You could try buying fruit and vegetables loose. Buying items to order from the deli counter saves packaging and in most cases also saves money. Similarly, buying items in bulk or buying refills or concentrated products reduces the amount of packaging waste that is produced and at the same time saves money. Some products have more packaging than other equivalents, and of course there are the criminally over-packaged products such as Easter eggs and other cases of poor practice. In general, vastly over-packaged goods can easily be avoided. By thinking before you buy, and choosing the right products, you will be using market forces to inform supermarkets and producers that consumers prefer products without excess packaging.

Q Can I recycle drinks cartons?

A Juice and milk cartons or Tetrapaks are made from a complex mix of paper, plastic and aluminium contained within seven layers; the composition varies according to the product contained. For example, products that begin to degrade when exposed to light need packaging containing a high proportion of aluminium to prevent light from reaching them.

Not every local authority currently recycles Tetrapaks in the UK. The main reason is that separating the three different materials in them requires very expensive specialised equipment. Also in the UK our recycling target system is weight based. Tetrapaks are very light compared to materials such as glass or paper and a tonne of Tetrapaks would take considerably more effort to collect and more space to store than a tonne of glass or paper. Local authorities have to meet demanding recycling targets under tight budgets and in the interest of the tax-paying public focus on materials that will enable them to meet their targets at the lowest cost.

Tetrapaks are however very recyclable. They are made up of at least 70 per cent paperboard which can easily be recycled and the remainder of the carton can be used as fuel in energy-from-waste facilities. Germany and Belgium have established Tetrapak recycling systems in place, and currently recycle as much as 65–70 per cent of their Tetrapaks.

Many UK local authorities are widening their recycling services to provide Tetrapak recycling. The cartons are collected and taken to a recycling plant in Sweden. However a plant is planned for the UK when services develop and a sufficient volume of Tetrapaks is available.

Q Is it true that green glass has to go to Argentina to be recycled?

A The UK imports twice as much green glass as it manufactures, mainly in the form of wine bottles. Most of the glass collected for recycling in the UK is green glass, and because the glass industry can't use it all, some surplus glass has been exported to glass reprocessors in countries such as Argentina.

There are various other types of glass that cannot be used in the UK packaging industry. These include batches of glass contaminated with non-recyclable glass i.e. heat-resistant glass, flat glass such as windows, and glass screens from discarded televisions. Over recent years significant developments in glass recycling have enabled the UK glass industry to make more use of glass that has previously been exported or sent to landfill. Glass that cannot be used for making new glass can be put to use in aggregates for the building industry, sandblasting systems, and a road-laying material called glasphalt. Other new developments use recovered glass in water filters, bricks, cement and concrete, sports surfaces and fibreglass insulation. Green glass does not have to go to Argentina or anywhere else to be recycled; there are now plenty of innovative opportunities for reusing or recycling it in the UK.

Q Why can't I recycle plastics?

A Plastics have many uses – from food packaging to car bumpers. There are around 50 different groups of plastics with over 100 different varieties. Plastic consumption is growing by about four per cent every year in western Europe and the world's annual consumption of plastic materials has increased from around five million tonnes in the 1950s to nearly 100 million tonnes today. Plastics make up around seven per cent of the average household dustbin contents, and the safe treatment of post-consumer plastics are a major source of concern.

The main reason why it is difficult to recycle plastics is that there are over 100 different varieties of plastic in circulation. Firstly, collecting and separating so many different types of plastic is a very complicated and time-consuming process because many types of plastics are blends of different polymers that can be difficult to identify. Secondly, post-consumer plastics tend to be heavily contaminated with food and other materials, which can cause problems with sorting and reprocessing. Thirdly, plastics tend to be very light: one tonne of plastics is equivalent to 20,000 two-litre drinks bottles or 120,000 carrier bags. Local authorities in the UK are required to meet demanding weight-based recycling targets within tight budgets. Collecting and storing one tonne of plastics instead of a heavier material such as glass costs them considerably more, and may prevent the council meeting its recycling targets. In some European countries local authorities collect many household waste plastics; however, due to plastic's high calorific value, much goes for incineration for energy generation.

Most local councils in the UK now provide services for the collection and recycling of plastic bottles. Plastic bottles are usually made out of either PET (Polyethylene terephthalate), HDPE (High-density polyethylene) and PVC (Polyvinyl chloride). Plastic bottles have been singled out because identifying and separating the polymer types is relatively easy compared to other plastics. They also make up 40 per cent of all plastics in the household waste stream, and so by targeting them, local authorities can potentially divert a large proportion of plastics from landfill. Plastic bottles are now collected and recycled in most areas in the UK, and 20 per cent of household plastic bottles were recycled in 2006. To reduce plastic, buy less! Buy juice and milk in glass bottles that are recycled or reused and drink tap water rather than expensive bottled varieties.

Lots of products are made from recycled plastic. They include polyethylene bin liners and carrier bags, PVC sewer pipes, flooring and window frames, building insulation board, video, DVD and CD cases, fencing and garden furniture, water butts, garden sheds and composters, seed trays, anoraks and fleeces, fibre filling for sleeping bags and duvets, and a variety of office accessories. If you buy products containing recycled plastic you will help the plastic recycling industry to develop and increase the opportunities for recycling more plastic.

Q Is it necessary to rinse recyclables, washing leftover products into the water system? Surely recycling plants wash the waste anyway?

A Containers should always be rinsed before they are put in the recycling, mainly for health and safety reasons. Even in the most automated materials recycling facility, operators transport the recyclables around the site, help sort the materials, and ensure that the facility runs smoothly. If containers are not rinsed before they are put into the recycling, the food or liquid inside begins to go off very quickly. By the time the recyclables reach the materials recycling facility the leftover liquid or food contained in them will have begun to go off, exposing staff to potentially harmful bacteria.

Rinsing recyclables also helps to ensure the machinery at the materials recycling facility sorts the materials accurately. Materials are sorted according to their properties, such as weight – plastic bottles being light and glass heavy, or by using magnets to sort steel cans from aluminium. If recyclables are contaminated with food or liquid the machinery may not be able to distinguish between different materials and will not be able to sort them properly. In addition, excess food and liquid contamination could transfer from the recyclables to the machinery and cause malfunction and possible breakdown of the facility.

Rinsing out is easy and there is no need to waste hot water in the process. You can rinse them at the end of the washing up, or put them in the dishwasher. By rinsing your recyclables you will help ensure that the sorting and transportation of recyclables is carried out efficiently, and make them safer and more pleasant for the operatives handling them.

Q Why should we reduce our use of plastic bags?

A It is estimated that over one million plastic bags are used globally each minute! Plastic bags are popular because they are light, versatile, waterproof and cheap to produce. The number of plastic bags we get through is staggering and the environmental costs and the impact of their production and disposal is increasing as more are used each year. Most plastic bags are only used once, for about 20 minutes, before being discarded.

Plastic bags contribute to litter – caught in trees, bushes and fences they make the built environment unsightly. When buried in a landfill site plastic bags do not break down for many years. It is estimated that plastic bags take around 500 years to degrade, although no one knows for sure because the bags haven't been around long enough to test the theory. Even then they do not fully degrade, they just break down to smaller and smaller parts that can end up in the food chain.

Discarded plastic bags also pose a severe threat to wildlife. Birds and other animals can get caught up in or choke on discarded plastic bags. Plastic bags that end up floating in the seas are often eaten by turtles mistaking them for jellyfish, with fatal consequences.

Plastic bags can accumulate in drains and sewers causing flooding and the spread of water-borne diseases due to blockages. Bangladesh banned plastic bags in its capital city Dhaka in 2002, after plastic bags blocked up drains leading to the prolonging of flooding in 1989 and 1998.

The manufacture of plastic bags wastes valuable resources. Plastic bags are made from oil, which is a finite and increasingly scarce resource. The extraction of the raw materials, along with the production and transportation of the bags, use energy and make a significant contribution to carbon emissions. Reducing the demand for plastic bags will reduce the environmental impact of their production and decrease the threat they pose to wildlife.

The Chinese government has recently introduced a ban on certain types of plastic bags and is planning to introduce a tax on others in a bid to conserve valuable resources and protect the environment.

In 2002, Ireland was the first country in the world to introduce a tax on plastic bags. Shoppers were required to pay 15 cents for every plastic bag they used. The tax reduced plastic bag use from 328 per person per year to only 21. After five years the effect of the tax reduced, and so it was raised to 22 cents in July 2007. The money is put into a fund for environmental projects. It is claimed that the tax has reduced plastic bag litter by more than 95 per cent. Scotland and England are currently considering introducing a plastic bag tax or ban and may be followed shortly by other European countries. Some UK towns have already taken the initiative and banned bags in their own community.

You can easily cut out your need for plastic bags by buying a reusable shopping bag. Many different types are available. Jute or cloth bags are widely available and last longer than the reusable plastic alternatives on offer in supermarkets.

To dispose of the plastic bags you have, many local authorities will collect them for recycling and there are now recycling banks at most supermarkets. By refusing the plastic bags offered and instead using your own long-life shopping bags, you can make a real reduction in the number of plastic bags consumed.

Bank Balance

Triodos is a bank with a unique approach to money, connecting savers and investors who believe in a more sustainable society with the organisations making it happen. It works like any other bank, but with one crucial difference: Triodos only finances organisations working to benefit people and the planet. What an enterprise sets out to do, and what motivates the people behind it, are the Bank's first consideration, examined before a loan's financial viability is even considered.

The projects that do make the grade are in areas ranging from renewable energy and recycling to Fairtrade and organic farming. Some are national names, including Cafédirect and Hugh Fearnley-Whittingstall's River Cottage, while the impact of others is felt more locally, like community groups and village shops. There's even an eco-publisher or two in there. A loan from Triodos Bank helped Alastair Sawday's dream of having one of the UK's most environmentally friendly offices become a reality in 2006 – with super-insulation, underfloor heating, a wood-pellet boiler, solar panels and a rainwater tank.

In this chapter the Triodos team answer a few common questions about money, banks and the ethical alternatives available.

William Ferguson
Communications Officer

Triodos ⚫ Bank

Q Is money the root of all evil?

A Despite some bad biblical press, money is neither good nor evil. It is more accurate to view money not as an ambivalent force in itself, but one that can be used for good or ill depending on what we choose to do with it. Use it wisely and it could provide solutions to some of the most daunting social and environmental problems facing us today.

Money's bad reputation stems in part from one of the Bible's most commonly misquoted passages, Timothy 6:10: "The *love of* money is the root of all evil". Reintroducing the often omitted words shifts the emphasis away from money, and on to us. Or rather to our role within a global financial system that focuses on the love of money and short-term gain.

Modern society has reduced money to numbers on a page, almost a commodity with a pure numeric meaning. Newspapers list reams of percentages earned or paid for money that can be borrowed, saved or invested. We are encouraged by the media, and conventional wisdom, to chase the highest interest rates. But it is surprisingly difficult to find information about how that money is used once we have deposited it, or who might be paying the price for these sometimes unfeasibly high rates of interest. The result is a culture of detachment from the impact of our financial decisions. By entrusting money to banks, many of us wash our hands of it and its consequences.

It is harder still to find a discussion about how money will make you feel, and how it can be integrated into your life. This is strange, given that money cannot be enjoyed or consumed. What can make us feel good about it is when it is used for purposes that we believe in and support – helping people out, rewarding work, educating and unlocking potential. To feel really prosperous, our financial choices should be in line with our long-term plans, with our deepest sense of values and fulfilment. To use money sustainably, we need to consider it as an extension of ourselves, which we can direct for the purposes that we choose.

While its misuse may have terrible consequences, money is not the root of all evil. Even the love of money in itself is not bad, or at least loving the benefits it can bring to people and the planet. Money is, at least in part, behind the success of millions of good ideas and progressive initiatives – from developing new sources of green energy and reducing our dependence on fossil fuels, to this book. What is most important is that we take a conscious approach to our money, consider its impact and whether the return we earn on it is worth the true cost. Ensure your financial choices are in line with the values you live by. Love your money, but for all the right reasons.

Q Aren't all banks as bad as each other?

A With the financial news agenda dominated by stories of banks overcharging, misleading and mistreating customers – coupled with announcements of multi-billion pound profits – you could be forgiven for thinking that all banks are as bad as each other. Of course, some of banking's bad reputation is well deserved. The fact that the industry's regulator, the Financial Services Authority, has recently introduced an initiative called 'Treating Customers Fairly' arguably says it all. It is odd that a sense of fair play has to be imposed by a regulator, and is not just business as usual.

But the concept of banking itself is not necessarily bad. Anyone who has bought a house with the support of a mortgage, or profited from savings and investments, can tell you that. Taken at its most basic level, banking is about human relationships. People save with a bank, and get a secure home for their money and interest in return. The bank lends this money to someone else for a higher rate of interest and profits from the difference. In essence, a bank is an organisation that brings two groups of people together: someone who has money to spare, and someone else who needs it. The bank can then put its expertise and experience into the mix, acting as facilitator and advisor.

The trouble with many modern banking relationships is that most of this humanity has been squeezed out. Many people do not consider it, but if you have savings in a bank, you are effectively lending money to it. Most banks are then free to profit from lending that money in anyway they see fit. It can be almost impossible to find out who they lend it to. The organisations they finance may be doing no harm, but equally could actively damage communities and the environment – by financing arms projects or colossal oil pipelines, for instance.

There are alternatives. Bangladeshi microfinance institution Grameen Bank and its founder Muhammad Yunus were jointly awarded the Nobel Prize in 2006 for providing basic financial services to people in the developing world. Closer to home, Local Exchange Trading Systems (LETS) are helping people to exchange all kinds of goods and services with one another, without the need for money. And a handful of ethical banks are breaking the financial mould, with a more progressive approach that values people and planet as well as profit. These social lenders do not just avoid investment in industries like the arms trade and nuclear power, but actively support positive social or environmental developments. Triodos Bank, for example, only lends money to organisations whose work benefits people and the planet. And, uniquely for a UK commercial bank, Triodos also publishes details of all the people it lends money to. This transparent approach means Triodos Bank's savers and investors can see exactly where their money is working.

This awakening of the connection between a bank's savers and its borrowers helps to move banking back towards its basic principles and to reintroduce human relationships. It is a better way of banking, and one that is increasingly popular.

As its influence grows, 'banking for good' could help make the world less like it is, and more like it should be.

Q Why don't more people bank ethically?

A Ethical consumption has never been so popular. People are choosing to spend more money than ever on products from ethical industries like organics and Fairtrade, and retailers are eager to profit from consumers' appetite for them. But despite record levels of spending ethically, many green consumers still fail to choose to save sustainably. Given the power and influence money has on the world, it is perhaps surprising that this paradox exists.

Why do people who live otherwise ethical lives choose to deposit their money with banks they recognise as having policies that can harm the planet? In 2007 Triodos Bank commissioned research to get to the bottom of this question. The report by Professor Alex Gardner – 'How Green is Your Money?' – studied people who regularly recycle and are happy to pay more for ethical products, but ignore their basic values when it comes to their banking choices. Participants in the study blamed a lack of services, inadequate information, and apathy for their decision not to do it. One factor they said they were most interested in is financial return – not where the money went. So why do they apply this ruthless criterion to their financial decisions, when they are prepared to pay more for ethical products elsewhere?

Maybe it is because banking is not yet dinner party conversation. Friends will often debate the merits of organic food and Fairtrade round the table, but we don't tend to discuss money matters, even with close friends. We are unlikely to know what they earn, what they have got saved and who they bank with. So there is little chance of picking up on the real stories behind banking – both the negative ones, and the uplifting projects that reflect the sort of changes we support through ethical shopping.

Participants in the study identified apathy as one of the factors holding them back. Because people don't make financial decisions on a regular basis – it is said that most people are more likely to get divorced than switch banks – they don't often have to think about their consequences. In contrast, every time they go shopping they are forced to make a choice about what to buy – especially when food and clothes are so visible. There is always the risk of bumping into a friend at the checkout with a basket full of the 'wrong' products.

Gemma Thorogood from Bristol typifies the ethical consumers interviewed in the study. "What I put in my shopping basket is important to me," she says. "When I can afford to, I buy my food from ethical suppliers – partly because I know the story behind organics and Fairtrade." She also suggests, "perhaps if there was more information about what the mainstream banks are doing I would have looked at the alternatives in more detail".

So what is the solution? We need a reassessment of how we think about the value of money, one that looks at finance in a more holistic way. This in turn could help to change attitudes and allow money to become a talking point. Only then can the public buy into the story as well as the products – only then can ethical money start to enjoy the same profile that has helped propel sales of organic and Fairtrade goods.

Q How can I use money as a force for good?

A According to popular myth, money makes the world go round. It is one of the most powerful ways we have in the developed countries of influencing the world around us. Every time we spend or save we are making a choice with wider repercussions than we might imagine. In a global economy where corporations cross international boundaries and can have higher GDPs than some countries, how we spend our money is a form of daily democracy. Buying Fairtrade coffee, for example, supports producers in the developing world. At the same time it sends a strong message to conventional coffee brands that their working practices are unacceptable. But more importantly it also shows the coffee industry that consumers are willing to pay more for socially responsible products. It's both carrot and stick in an area that almost all mainstream businesses are most interested in – their profit margins.

The beauty with money is that it only takes a few small changes to start to make a big difference. The most immediate thing you can do is think about what your choice means to society and the environment. If you need to buy something, consider the product you purchase and what message it sends out to retailers and manufacturers. If you can help to create sufficient demand for sustainable products, the market will supply them. Also consider what you do with your money when you're not spending it. Your savings and investments help fuel the global economy, so make sure your choice of bank matches your wider approach to the world.

But beware of 'greenwash'. As big business has started to see the potential profits in ever-expanding ethical consumption, many have promised sustainability, without backing their claims with significant changes to the way they work. Which means you have to be well informed to ensure you are really making a difference, not just having the wool pulled over your eyes. Whether it is your choice of coffee, clothing or bank, question your supplier's claims. And if you don't feel happy with their answers, use your influence as a consumer they depend on and go elsewhere.

You have the power to change things. Your money can be a compelling force for good. If money does make the world go round, make sure you are helping turn it in the right direction.

White Wash

For over 27 years, Ecover has been devoted to developing and producing effective and ecological washing and cleaning products made from plant and mineral based ingredients. Ecover's aim is to provide effective sustainable alternatives for washing and cleaning that can be used daily by people all around the world.

Ecover's ecological principles extend far beyond the products they create. Ecover's vision of sustainability takes ecological, economic and social aspects into account from the origins of the raw materials, to the complete biodegradation of the final products. Strict criteria are employed along the way as guidelines for all business operations. The products themselves are manufactured in Ecover's unique, world-famous ecological factory.

Ecover is a company that operates with sustainability at its very core. Ecover is constantly innovating and pushing boundaries to create new and more effective products that have minimum negative impact on the environment.

Mick Bremans
Managing Director

ECOVER
Effective Cleaning - Ecologically

Q Doesn't industry account for most water usage? Why should I bother saving any?

A Without a doubt, industry accounts for a huge amount of the water used and wasted in the UK. Envirowise, the government-funded programme tasked with offering advice and environmental consultation for UK businesses, has estimated that the UK industry as a whole uses a total of 1.3 billion cubic metres of water every year – a figure that could be reduced by 30 per cent. This not only has a significant environmental impact but also hits many businesses' bottom line. Introducing water-saving measures in industry could offer a national cost saving of £507 million.

However a great deal of water is also wasted in the home – probably more than you would think. The average person in the UK uses 50 litres per day, 50 per cent more water per person than 25 years ago. The majority of this water is used in simple everyday tasks like brushing our teeth, washing our clothes and flushing the toilet.

Consider these statistics:
- A huge 30 per cent of our household water goes on flushing the loo. The average family uses the equivalent of two baths of water per day just for this!
- Leaving the tap on while you brush your teeth wastes up to five litres a minute. If the entire population of the UK turned the tap off we could save a total of 180 million litres of water a day: enough to supply 500,000 houses a day.
- A shower uses only 60 per cent of the water used for a bath.

These statistics prove that even little tasks can drain our water resources. We all have a responsibility to make an effort to reduce our impact on the water system and take pride in our individual actions. Following these simple tips will help you reduce your own water use and impact on the environment.

- Install a professional water-saving device in your toilet or simply fill a one litre bottle of water, screw on the lid and place in the water cistern. Even better, use rain water to flush the toilet!
- Fix any dripping taps, they can waste up to four litres of water a day.
- Wash vegetables and fruit in a bowl rather than under a running tap.
- Always run washing machines and dishwashers with a full load and on the economy setting.

If the whole of the UK made these small changes we could make a big difference.

Some industries are also trying to do their bit to save water, and many action groups like Envirowise have been established to help businesses to reduce their water consumption. For example, Ecover, producer of ecological washing and cleaning products, runs a sustainable business from two 'ecological' factories in Belgium and northern France. Both factories are made from recycled and recyclable materials and are energy efficient. At the factory in Belgium, waste water is processed and cleaned on site and returned to its original state of natural cleanliness. The factory in France has the foundations in place to carry out this process too, once the production impact has been assessed. Machinery is cleaned using air pressure, rather than water, to save on resources. The company constantly tries to innovate and minimise its impact on the environment. According to their philosophy, what is good enough for today is never good enough for tomorrow.

Q Why does it matter what I put in the water system? It goes through water purification plants before it comes out of my tap again.

A The UK is fortunate to have an efficient and well-developed water system so when we turn our taps on for a drink or to flush the toilet there is always enough water to do the job. To make this happen the water industry supplies around 15 billion litres of water per day to the population of England and Wales. It also collects and disposes of over ten million tonnes of waste water every day.

As a result the water industry has the potential to have a great impact on the environment. We are completely dependent on water, but the water system is very fragile. All of our water starts as rain, which we collect from rivers and underground wells, but we take more than the system can bear and then we waste it. We all have a duty to lighten the burden on our water system by trying to keep it pollution free.

Under our kitchen sinks there are a number of products that can cause irreversible damage to our water system. They impact on our system in two ways: by creating an unnecessary burden as more water is required to neutralise their impact, and by polluting our water supply.

Chemicals, including everyday items like cleaning products and washing powder, can all harm our systems beyond repair when they disappear down the plughole. Ecological cleaning products are now available for people who want to minimise their impact on the environment. These are usually based on plant and mineral ingredients derived from renewable sources and biodegrade quickly and completely after use.

Critical dilution volumes offer a way for us to understand the burden of cleaning products and toxic chemicals on the water system. They show how much water it takes to neutralise the chemicals contained within a product. Consider these statistics from Ecover:

- Ecover ecological toilet cleaner can take 400 times less water to neutralise than non-ecological products.
- For every wash using non-ecological washing powder it takes 10–12,000 litres of water to treat it until it can re-enter our water system safely. Ecover biological washing powder can take less than 4000 litres of water for every wash.

It is also important to consider the impact of chemicals on the aquatic ecosystem. Sewage from towns and industry is processed and pumped into the rivers and the sea. The ingredients in conventional cleaning products, paints and other chemical products cannot be removed efficiently and do not biodegrade so pass directly into the aquatic ecosystem. Sea-life becomes contaminated and the toxins build up and come back to us in the food chain through the seafood we consume.

We can all try to reduce our burden on the water system by following a few simple rules:

- Check all your plumbing. Some sinks in older houses and most outside drains are not connected to the sewer, but discharge untreated into surface drains and then into rivers and streams.
- Never put any oils or chemicals down the drain.
- Be aware that leaking bin bags, animal waste and car oil can also be washed into drains by the rain.
- Avoid all pesticides and garden chemicals. Use natural ecological products where possible.

Q I'm washing my clothes at lower temperatures now, isn't that enough?

A Washing clothes at lower temperatures is important when you consider that up to 90 per cent of the energy used in washing goes on heating the water. However, this is just a starting point: there are many more issues in the environmental equation.

Firstly, ask yourself if you actually need to wash the garment: sometimes it's just habit to take things off and put them in the wash basket. Outer layers of clothing can be worn more than once without laundering.

You can also minimise your environmental impact by making sure you use your washing machine effectively. Washing machines work most efficiently when they are full, but it is important to keep a fist's breadth of free space at the top of the washing in the drum. Simple things like remembering to use the half load button will ensure you use less energy, water and detergent to get the job done.

If your washing machine is on its last legs, and it really can't be repaired, replace it with an 'A' rated energy-saving machine. Statistics from the Energy Saving Trust show that by making the switch you could cut your energy consumption by up to a third.

Finally, it's very important to consider what you use to wash your clothes. Conventional laundry products are based on petrochemicals and chemicals which are non-renewable and damage the ecosystem. They also contain optical brighteners – chemicals that reflect light, making clothes look brighter than they are. Optical brighteners form an irreversible bond with the skin and add nothing to the cleaning performance of the product.

All these factors should be considered in conjunction when washing clothes to minimise environmental impact.

Alternative products available include plant and mineral based ecological cleaners, natural soap nuts and laundry balls filled with ionised particles that produce ionised oxygen, penetrating clothing fibres to clean them.

Q Surely there are much more significant climate change issues to worry about than water?

A Environmental issues are highly complex and it's sometimes difficult and confusing to know what you can do to really make a difference.

Climate change is the major issue facing us at the moment, but the greenhouse gas carbon dioxide (CO_2) is only one part the equation. The world is totally dependent on water, but like many of the Earth's natural commodities it is something westerners expect to have in abundance and often take for granted as they turn on the tap.

While water quality has improved dramatically in the developed countries, the burden of phosphates and nitrates is still high. These chemicals travel from rivers into the sea, polluting sea-life. The toxins then come back to us through the food chain.

The severe floods in Gloucestershire and Yorkshire in 2007 make it difficult for us to imagine ever suffering from a significant drought, but the issues extended far beyond flooded homes. Thousands of people were left without water to drink, wash with, or perform many of the everyday functions we take for granted.

In England and Wales our water system provides 1334 cubic metres (m^3) of water per person a year. This is much less than France, which has nearly double at 3065m^3. Even the hotter Mediterranean countries of Italy and Spain have a larger allowance than the UK. South-east

England has even less water per person due to its high population density and low rainfall. The Thames Valley has only 266m³, one-fifth of the England and Wales average.

We all have a responsibility to lessen the burden on our water system by saving water where possible, and making sure we are not unnecessarily polluting it. Water shortages don't just affect us: they can also seriously harm our environment. Our water comes from rivers and groundwater so every drop we use has a direct effect on these ecosystems. Fish, wetland birds and other wildlife that rely on ponds, rivers and streams struggle to survive when these dry up or run low. Sources of food and breeding sites for wildlife can be lost and fish can die through lack of oxygen.

It is easy to look at these issues on a purely national level in isolation to the rest of the world, but the water cycle is a complex, dynamic system. Two-thirds of our planet is made up of water so issues of water consumption and pollution link us all on a global scale. Across the world 1.1 billion people (roughly one-sixth of the world's population) still do not have access to safe water. The charity WaterAid works on projects in developing countries to provide water, sanitation and hygiene education. As inhabitants of a country with a developed water system we have a duty to respect the water supply we have.

The effect of rising greenhouse gas concentrations

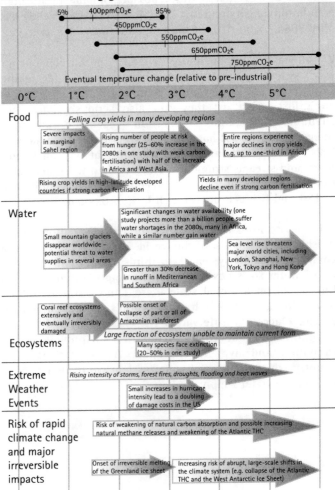

This table from the Stern Review relates greenhouse gas concentrations, global temperatures and consequences. The bars at the top show the probability that a temperature will not be exceeded for a given concentration, i.e., on the 450ppmCO$_2$e line there is only a 50% chance (marked with a vertical line) that temperatures will not exceed 2°C, but a 95% chance they will not rise above 3.8°C. The described consequences also show probability: from slight on the left to very likely on the right.

Resources

Websites

Carbon Addicts
Cap and Share
www.capandshare.org
Carbon Calculator
www.carboncalculator.co.uk
Carbon Footprint
www.carbonfootprint.com
Centre for Alternative Technology
(CAT) www.zerocarbonbritain.com
C S Kiang tinyurl.com/4vb75m
Feasta www.feasta.org
IPCC www.ipcc.ch
NASA www.giss.nasa.gov
Oceana www.oceana.org
One Planet Living
www.oneplanetliving.org
Oxford Research Group
www.oxfordresearchgroup.org.uk
Pew Centre www.pewclimate.org
Rio Earth Summit, Severn Suzuki
tinyurl.com/2qd47o
SCAD
www.salt-of-the-earth.org.uk
Tyndall Centre www.tyndall.ac.uk
UK Met Office Hadley Centre
www.metoffice.gov.uk
Union of Concerned Scientists
www.ucsusa.org
Uranium Information Centre
www.uic.com.au
US National Academy of Sciences
www.nas.edu
US National Oceanic and
Atmospheric Administration
www.noaa.org
Vostock ice core data
tinyurl.com/4zf2nz
Worldwatch Institute
www.worldwatch.org

Energy Junkies
British Wind Energy Association
www.bwea.com
Campaign for Better Transport
www.bettertransport.org.uk
CAT www.cat.org.uk
Energy Saving Trust
www.est.org.uk
National Energy Foundation
www.nef.org.uk
Solar Trade Association Ltd
www.solartradeassociation.org.uk

Travel Turbulence
Ed Gillespie
www.lowcarbontravel.com
George Monbiot
www.monbiot.com
Go Slow England blog
goslowengland.wordpress.com
Green Traveller
www.greentraveller.co.uk
Man in Seat 61 www.seat61.com
Plane Stupid
www.planestupid.com
Sawdays www.sawdays.co.uk
Slow Planet www.slowplanet.com
Sustrans www.sustrans.org.uk

Funny Farm
Compassion in World Farming
www.ciwf.org.uk
David Suzuki Foundation
www.davidsuzuki.org
GRAIN www.grain.org
Louis Bolk Institute
www.louisbolk.org
Newcastle University
www.ncl.ac.uk
Soil Association
www.soilassociation.org
UN FAO Organic Agriculture
www.fao.org/organicag

Wasting Away

Ban Waste www.banwaste.org.uk
Plastic Bag Free
www.plasticbagfree.com
Resources not Waste
www.resourcesnotwaste.org
Waste Watch
www.wastewatch.org.uk
WRAP www.wrap.org.uk
Zero Waste Alliance
www.zwallianceuk.org

Bank Balance

BankTrack www.banktrack.org
Ethical Pulse
www.ethical-junction.org
Grameen Bank, Bangladesh
www.grameen-info.org
Money Facts
www.moneyfacts.co.uk
Transition Towns
www.transitiontowns.org
Triodos Bank www.triodos.co.uk

White Wash

Ecover www.ecover.com
The Environment Agency
www.environment-agency.gov.uk
Envirowise
www.envirowise.gov.uk
WaterAid www.wateraid.org/uk

Bibliography

Ed Ayres, *God's Last Offer*, 1999
E B Balfour, *The Living Soil*, 1943,
reprinted 2006
Michael J Benton, *When Life
Nearly Died*, 2003
James Bruges, *The Big Earth
Book*, 2007
Peter Bunyard, *Gaia, Climate and
the Amazon*, 2005
Cabinet Office, *The Stern
Review*, 2006
CAT, *Zero Carbon Britain*, 2007
Roger Deakin, *Waterlog*, 1999
Richard Douthwaite, *The Growth
Illusion*, 1999
Meyer Hillman, *How We Can Save
The Planet*, 2004
Kummeling et al., *Consumption of
organic foods and risk...*, 2007
Larry Lohmann, *Carbon
Trading*, 2006
James Lovelock, *The Revenge of
Gaia*, 2006
Mark Lynas, *High Tide*, 2004
Aubrey Meyer, *Contraction and
Convergence*, 2000
New Economics Foundation,
Up in Smoke, series from 2004
Martin Rees, *Our Final
Century*, 2003
Rist et al., 'Organic diet and
breast milk', *British Journal of
Nutrition*, 2007
Andrew Simms, *An Environmental
War Economy*, 2001
Peter Singer, *One World*, 2002
Swann, *Report of the Joint
Committee on the use of
Antibiotics in Animal
Husbandry*, 1969
Colin Tudge, *Sow Shall We
Reap*, 2004
UNDP, *Human Development
Reports*, annual
US-NAS, *Abrupt Climate
Change*, 2001

Index

Other Fragile Earth titles

One Planet Living by Pooran Desai & Paul King

£4.99   ISBN 978-1-901970-85-2
This inspiring little book shows how sustainable
living can be easy affordable and attractive.
Over 25,000 copies sold.

The Big Earth Book by James Bruges

£25   ISBN 978-1-901970-87-6
"A back-catalogue of core environmental thinking
from the past 30 years... Pictures that are more suited
to National Geographic and a layout so accessible
that your children will pick it up." *The Ecologist*

New titles published in autumn

Ban the Plastic Bag by Rebecca Hosking *et al*

Published October 2008
£4.99   ISBN 978-1-906136-16-1
This book is a call to action, entreating every village,
town and city in the country to ban the bag.

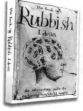

The Book of Rubbish Ideas by Tracey Smith

Published September 2008
£6.99   ISBN 978-1-906136-13-0
Full of practical tips and quirky ideas, this is the
essential guide to reducing household waste.